THE
BIG
ONES

THE
BIG
ONES

How Natural Disasters Have Shaped Us

(and What We Can Do About Them)

Dr. Lucy Jones

DOUBLEDAY

New York

All rights reserved. Published in the United States by Doubleday, a division of
Penguin Random House LLC, New York, and distributed in Canada by Random
House of Canada, a division of Penguin Random House Canada Limited, Toronto.

www.doubleday.com

DOUBLEDAY and the portrayal of an anchor with a dolphin are
registered trademarks of Penguin Random House LLC.

Book design by Michael Collica
Frontispiece by Frdric Lahme EyeEm
Jacket design by Emily Mahon
Jacket photograph © Joris Grling/EyeEm/Getty Images

Library of Congress Cataloging-in-Publication Data
Names: Jones, Lucile M., author.
Title: The big ones : how natural disasters have shaped us (and what we can
do about them) / by Lucy Jones.
Description: First edition. | New York : Doubleday, 2018. | Includes
bibliographical references.
Identifiers: LCCN 2017036796 | ISBN 9780385542708 (hardcover) |
ISBN 9780385542715 (ebook)
Subjects: LCSH: Natural disasters—History. | Natural disasters—Social
aspects. | BISAC: NATURE / Natural Disasters. | SCIENCE / Earth Sciences /
Seismology & Volcanism. | HISTORY / Social History.
Classification: LCC GB5014 .J66 2018 | DDC 363.34—dc23
LC record available at https://lccn.loc.gov/2017036796

MANUFACTURED IN THE UNITED STATES OF AMERICA

2 4 6 8 10 9 7 5 3 1

First Edition

For our unsung heroes: the city planners, building officials, and others who love their communities and work every day to prevent future natural disasters from becoming human catastrophes

CONTENTS

THE
BIG
ONES

IMAGINE AMERICA WITHOUT LOS ANGELES

Earthquakes are happening constantly around the world. The seismic network that measures earthquakes in Southern California, where I live and spent my career as a seismologist, has an alarm built into it that goes off if no earthquake has been recorded for twelve hours—because that must mean there's a malfunction in the recording system. Since the network was put into effect in the 1990s, Southern California has never gone more than twelve hours without an earthquake.

The smallest earthquakes are the most common. Magnitude 2s are so small they are felt only if someone is very nearby their epicenter, and one happens somewhere in the world every minute. Magnitude 5s are big enough to throw objects off shelves and damage some buildings; most days a few of these strike somewhere. The magnitude 7s, which can destroy a city, occur more than once a month on average, but luckily for humanity, most take place underwater, and even those on land are often far from people.

But for more than three hundred years, none of these, not even the tiniest, has occurred on the southernmost part of the San Andreas Fault.

Someday that will change. Big earthquakes have happened on the southern San Andreas in the past. Plate tectonics hasn't suddenly stopped; it is still pushing Los Angeles toward San Francisco at the same rate your fingernails grow—almost two inches each year. Even though the two cities are in the same state and on the same continent, they are on different tectonic plates. Los Angeles is on the Pacific plate, the largest of the world's tectonic plates, stretching from California to Japan, from the Aleutian Arc of Alaska to New Zealand. San Francisco is on the North American plate, which extends east to the Mid-Atlantic Ridge and Iceland. The boundary between them is the San Andreas Fault. It is there that the two plates get carried slowly past each other; their motion cannot be stopped any more than we could turn off the sun.

In a strange paradox, the San Andreas produces *only* big earthquakes because it is what seismologists consider a "weak" fault. It has been ground so smooth, across millions of years of earthquakes, that it no longer has rough spots to stop a rupture from continuing to slip.

To understand the mechanics of it, imagine you've laid a large rug on the floor of a room that has wall-to-wall carpeting. After placing it, you decide that, on second thought, you want to move it one foot closer to the fireplace. If it had been laid on a hardwood floor, it would be easy enough to move: you could simply grab the side nearer to the fireplace and pull. But it's on carpeting, so the friction between the carpet and the rug makes that impossible. What could you do? You could go to the far side of the rug, pick it up off the carpeting, and put the edge of the rug where you want it, a foot closer to the fireplace. You now have a big ripple, which you could push across the rug until you've reached the end, at which point the entire rug would be one foot closer to the fireplace.

In an earthquake, a seismologist sees not a ripple but a *rupture front.* The motion of that ripple across the "rug" of the San Andreas Fault creates the seismic energy that we experience as an earthquake. It is a *temporary local reduction in friction,* allowing a fault to move at *lower stress.* In the same way that the rug couldn't move all at once, an earthquake too must begin at one particular spot on its surface, its epicenter, and the ripple must roll across it for some distance.

The distance the rupture front travels is one of the chief determinants of an earthquake's size. If it moves a yard and stops, it is a magnitude 1.5 earthquake, too small to be felt. If it goes for a mile down the fault and stops, it's a magnitude 5, causing a little damage nearby. If it goes on for a hundred miles, it is now a magnitude 7.5, causing widespread disruption.

The San Andreas Fault has been smoothed to such a degree that now, when an earthquake begins, there is nothing left to keep it small. The ripple will continue to move down the fault, radiating energy from each spot it crosses, creating an earthquake that lasts for a minute or more and a magnitude that grows to 7 or even 8. Only after such an earthquake has broken the fault and made new jagged edges can it begin to produce smaller, less damaging earthquakes.

So we wait for that big earthquake. And wait.

The southernmost part of the fault had its last earthquake sometime around 1680. We know this because it offset the edges of Lake Cahuilla, a prehistoric lake in much of what is now the Coachella Valley, filling with water the flats where the Coachella music festival meets each year. It left behind geologic markers, as did previous earthquakes, so we know that there were six earthquakes between AD 800 and 1700. That means the 330 years since the last earthquake on this part of the San Andreas is about twice the average time between its previous earthquakes. We don't know why we are seeing such a long interval. We just know that plate tectonics keeps on its slow, steady grind, accumulating more offset and energy to be released the next time. Since the last earthquake in Southern California, about twenty-six feet of relative motion has been built up, held in place by friction on the fault, waiting to be released in one great jolt.

Someday, maybe tomorrow, maybe in a decade, probably in the lifetimes of many people reading this book, some point on the fault will lose its frictional grip and start to move. Once it does, the weak fault, with all that stored energy, will have no way of holding it back. The rupture will run down the fault at two miles per second, its passage creating seismic waves that will pass through the earth to shake the megalopolis that is Southern California. Maybe we will be lucky and the fault will hit something that can stop it after only a hundred miles or so—a magnitude 7.5. Given how much energy is already stored, however, many seismologists think it will go at least two hundred miles, and thus register 7.8, or even 350 miles and reach 8.2.

If it ruptures as far as central California, all the way to the section of the fault near Paso Robles and San Luis Obispo, it will hit a part of the San Andreas that behaves differently. This part accumulates a fingernail-growth rate of tectonic offset, just like the rest of the fault. But it's what is known as a "creeping section." Instead of storing energy to release in one big earthquake, the energy here oozes in small motions, sometimes with little earthquakes, sometimes with no seismic energy at all. We think, we hope, that the creeping section will act as a pressure valve of sorts, keeping the earthquake from growing any bigger than 8.2.

*

In 2007–8, as science advisor for risk reduction at the U.S. Geological Survey, I led a team of more than three hundred experts in a project we called ShakeOut, to anticipate just what such an earthquake will be like. We created a model of an earthquake that moves across the southernmost two hundred miles of the San Andreas, extending from near the Mexican border to the mountains north of Los Angeles—a likely outcome, though still short of the worst-case scenario.

In the earthquake we modeled, we found that Los Angeles would experience intense shaking for fifty seconds (compare this to the seven seconds of the Northridge earthquake in 1994, which caused $40 billion of damage). A hundred other neighboring cities would as well. Thousands of landslides would cascade down the mountains, blocking our roads, burying houses and lifelines.

In our model, fifteen hundred hundred buildings collapsed and three hundred thousand were severely damaged. We know which ones. They are the types of buildings that have collapsed in other earthquakes in other locations, and which we no longer allow to be built. But we have not forced existing buildings to be retrofitted to accommodate what we know. We might see some high-rise buildings collapse. The 1994 earthquake in Los Angeles and the 1995 earthquake in Kobe, Japan, exposed a flaw in how steel buildings had been constructed, causing cracks in their steel frames. Those buildings are still standing in downtown Los Angeles. We are going to see many brand-new buildings "red-tagged," too dangerous to enter and in need of major repairs or demolition. Our building codes do not require developers to make buildings that can be *used* after a major earthquake, only buildings that don't kill you. If the code works as it is supposed to, about 10 percent of the new buildings constructed to the latest code will be red-tagged. Maybe 1 percent will have partial collapse. A 99 percent chance of not collapsing is great for one building, but accepting the collapse of 1 percent of the buildings in a city with a million buildings is a different matter. The earthquake will probably not kill you, but it will likely make it impossible for you to get to work—for a very long time.

Of the results we projected, one of the most frightening was the impact of fires triggered by the earthquake. Earthquakes damage gas lines; break electrical items and throw them onto flammable fabrics; spill dangerous chemicals; and generally have many, many ways of starting fires. Two of the biggest urban earthquakes of the twentieth century were the 1906 San Francisco and 1923 Tokyo (Kanto) earthquakes. Both set off fires that turned into firestorms and burned down much of those cities. Some people think that modern technology has solved much of the fire problem because the two big California earthquakes of the late twentieth century, the 1989 Loma Prieta earthquake in San Francisco and the 1994 Northridge earthquake in Los Angeles, did not lead to devastating fires. This is a mistake. Not because technology hasn't changed, but because, in the eyes of seismologists, Loma Prieta and Northridge were not big earthquakes. Those who lived through them may dis-agree, and the damage they inflicted on those cities is undeniable. But these people simply don't know what a really big earthquake will be like.

What seismologists call "great" earthquakes (magnitude 7.8 and larger) are not just about stronger shaking—they are also about much larger areas. Loma Prieta and Northridge caused their stron-gest shaking near their epicenters, neither of which was in an urban core. Loma Prieta's was in the Santa Cruz Mountains; the stron-gest shaking of Northridge was felt in the Santa Susana Mountains. Even so, more than a hundred significant fires broke out in each of those earthquakes. They were fought through mutual aid. San Francisco and Los Angeles put out calls for help, and firemen from other jurisdictions poured in to help. Citywide fires were averted because of the amazing, courageous work of firemen from across the region.

When an earthquake like the one we modeled happens, every city of Southern California will have fires that need to be fought. Calls for help will be answered with desperate pleas for help in return. Aid will have to come from Northern California, Arizona, and Nevada. Those firemen will have to come to Southern Califor-nia from the other side of the San Andreas Fault, which will have moved twenty to thirty feet, offsetting all the highways into the region. Those responders will struggle, maybe for days, to bring equipment across broken roads. The firemen who are here will be sent to fight fires in places where the pipes feeding the fire hydrants have broken and gone dry. Our analysis, reviewed by the fire chiefs who had led the firefighting in Northridge and Loma Prieta, con-cluded that the fires would double the losses of the earthquake, in terms of both economic impact and casualties. Sixteen hundred fires could break out, twelve hundred growing large enough to re-quire more than one fire company. We don't have that many fire companies in all of Southern California.

As bad as this picture looks, it could be worse. In ShakeOut,

I got to specify the weather. I made it a cool, calm day. Unfortunately, I don't get to do this for the real thing. If the earthquake happens during the infamous Santa Ana winds, which have spread great Southern California wildfires and caused billions of dollars in losses, the fires that get started may be unstoppable.

Most of us will survive. Our estimate was that eighteen hundred people will die and fifty-three thousand will need emergency medical care. A significant number of hospital beds will be out of commission as hospitals suffer their own damage. And it will be very difficult to get to them. Bridges will be impassable, collapsed buildings will leave rubble in the street, and power will be knocked out, darkening traffic lights. Many people will be trapped in buildings; first responders will be overwhelmed. Most victims will be rescued by their neighbors. Losses will exceed $200 billion.

Life will not return to any semblance of normality for quite some time for the residents of Southern California. In the following months, tens of thousands of aftershocks will occur, some of which will be damaging earthquakes in their own right. The systems that maintain urban life—electricity, gas, communication, water, and sewers—will all be broken. The transport systems that bring food, water, and energy into the region all cross the San Andreas and will be cut. In a simpler world, when you lose your sewer system, you build a temporary outhouse in the backyard. In the dense urban environment of a modern city, a lack of sewers is a potentially catastrophic public health crisis. Cities are possible because of the complex engineering systems that support life. Those will be lost in such an earthquake.

Half of the total financial losses in our model were from lost business. A beauty salon cannot reopen without water. Offices cannot function without electricity. Tech workers cannot telecommute without Internet capabilities. Retail stores struggle if their clerks and customers don't have the means of transportation to get there. Gas stations cannot pump gas without electricity and cannot take your credit card if they're not online. And how many of us will want to stay in Los Angeles, much less go to work, when none of us have had a shower in a month?

Here we reach the limit of our technical analysis. Our scientists and engineers and public health experts can estimate buildings down, pipes damaged, legs broken, transportation disrupted. But the future of Southern California is the future of communities. We know what will happen to its physical structure, but what will happen to its spirit?

*

Natural disasters have plagued humanity throughout our existence. We plant farms near rivers and near the springs that form along faults, for their access to water; on the slopes created by volcanoes, for their fertile soil; on the coast, for fishing and trade. These locations put us at risk of disruptive natural forces. And indeed we are familiar with the occasional flood, tropical storm, passing tremor. We learn how to construct levees, perhaps a seawall. We add some bracing to our buildings. We are not quite so scared after the tenth minor quake. We begin to feel confident that we can control our natural world.

Natural hazards are an inevitable result of the earth's physical processes. They become natural disasters only when they occur within or near human construction that fails to withstand the sudden change they wreak. In 2011, a magnitude 6.2 earthquake occurred in Christchurch, New Zealand, killing 185 and causing roughly $20 billion in losses. Yet an earthquake of that size happens every couple of days somewhere in the world. This relatively minor earthquake became a disaster because it occurred right under the city, and the buildings and infrastructure were not built strong enough to withstand it. Natural hazards are inevitable; the disaster is not.

I have spent my professional life studying disasters. For much of my career, I was a researcher in statistical seismology, trying to find patterns and make sense of when and how earthquakes occur.

Scientifically, my colleagues and I could prove that compared to human timescales, earthquakes occurred randomly. But we found that "random" was an idea we could not convince the public to accept. So, recognizing that the desire for prediction was really a desire for control, I shifted my science toward predicting the *impact* of natural disasters. My goal was to empower people to make better choices—to prevent the damage from happening in the first place.

The U.S. Geological Survey, the government agency charged with providing the science about geological hazards, was my life-long professional home. In a pilot project in Southern California, and later for the nation, we studied floods, landslides, coastal erosion, earthquakes, tsunamis, wildfires, and volcanoes, with the objective of connecting communities to the scientific information that could make them safer, whether it was predicting landslides during rainstorms, recommending wildfire control in ecosystem management, or better judging our priorities when it comes to mitigating the risk of a big earthquake.

I was also one of the scientists who provided information to the public after earthquakes. I found people were desperate for science, but often not for the reason I expected. I saw the ways it could be used to halt the damage. But in times of natural disaster, the public turns to scientists to minimize not just destruction but also fear. When I gave the earthquake a name and a fault and a magnitude, I inadvertently found myself serving the same psychological function as priests and shamans have done for millennia. I was taking the random, awesome power of Mother Earth and making it look as though it could be controlled.

Natural disasters are spatially predictable—where they occur is not random. Floods happen near rivers, big earthquakes (generally) strike along big faults, volcanic eruptions take place at the site of existing volcanoes. But *when* they happen, especially compared to human timescales, is random. Scientists say an occurrence is "random about a rate." That means we know, in the very long term, how many of them take place. We know enough about a fault to know that earthquakes occur—have to occur—with a certain frequency. We can study a region's climate to the extent that its average rainfall becomes predictable. But whether this year brings floods or drought, whether the largest earthquake along the fault this year is a magnitude 4 or 8—that is purely random. And we humans don't like it. Random means every moment presents a risk, leaving us anxious.

Psychologists describe a "normalization bias," the human inability to see beyond ourselves, so that what we experience now or in our recent memory becomes our definition of what is possible. We think the common smaller events are all that we have to face, and that, because the biggest one isn't in anyone's memory, it isn't real. But in the earthquake that ruptures through the full length of a fault, the flood described as Noachian, the full eruption of a volcano, we see more than the common disaster. We face catastrophe.

In that catastrophe, we discover ourselves. Heroes are made. We laud the quick thinking, the unquenchable will to survive. We see extraordinary acts of courage committed by ordinary people, and we praise them for it. The firemen who run into a burning building when everyone else is running out hold a special place of honor in our society. Disaster movies always have as their hero the daring responder, from Charlton Heston in 1974's *Earthquake,* to Tommy Lee Jones in the 1997 film *Volcano,* to Dwayne "The Rock" Johnson in 2015's *San Andreas.* There is likewise a villain, in the public official who covers up the warning, or a selfish, scared victim who claims the last lifeboat for himself.

We show compassion for the victims, knowing that we could have been the one hit. Indeed, it is the randomness of the victimization that forms much of our emotional response, that encourages generous donations. For many people, helping the victims serves as a sort of unconscious good luck charm, warding off the same fate for themselves. We pray to God to protect us from the danger.

When the prayers fail and the catastrophe is upon us, we seem incapable of accepting that it is inexorably, infuriatingly random.

We turn to blame. For most of human history, the great disasters have been seen as a sign of the gods' displeasure. From the biblical Sodom and Gomorrah to the devastating earthquake of 1755 in Lisbon, those who survived, those who witnessed, declared that the victims were being punished for their sins. It allowed us to pretend that we could protect ourselves by not making the same mistakes— that we had no reason to fear the bolt out of the blue.

Modern science may have changed many of our beliefs, but it hasn't swayed our subconscious impulses. When that great Southern California earthquake finally strikes, I know two things will happen. First, rumors will spread that the scientists know that another earthquake is coming, but that we aren't saying anything to avoid scaring the public. This is the all-too-human rejection of the random, an attempt to form patterns, to find reassurance. Second, there will be blame. Some will blame FEMA, accusing them of a poor response. Some will blame the government for allowing bad buildings to have been constructed (maybe even the same people who fought against mandatory improvements of those weak buildings). Some will blame scientists for not listening to that week's earthquake predictor. Some, in a pattern we have seen for centuries, will blame the sinners of the hedonistic La-La Land.

The last thing any of us will want to do is accept that, sometimes, shift just happens.

Most cities have the potential for a Big One in their future. Those harbors, fertile fields, and rivers that make everyday life viable are there because of natural processes that can produce disasters. And that Big One will be qualitatively different from the smaller-scale disasters in our recent past. It is a disaster when your house is destroyed. It becomes a catastrophe when not just your home but your neighbors' homes and so much of your community's infrastructure are destroyed that societal functioning itself collapses. We have choices to make, right now, that could make our cities much more likely to survive and recover from these great natural disasters when they strike. We can make informed choices only if we consider our potential future, and if we take a hard look at our knowable past.

With this book I tell the stories of some of the earth's greatest catastrophes, and what they reveal about ourselves. Each was the Big One of its region, shifting the nature of that community. Together they show how our fear causes us to respond to random catastrophe—the reasoning we employ, the faith we manifest. We will see the limitations of human memory, which keeps us from believing that the one-in-a-million, or even the one-in-a-thousand, will ever affect us. And we will face the knowledge that our risk is growing. Because of the increasing density and complexity of our cities, more people than ever before are at a greater risk of losing the systems that maintain life.

We will come to a place where all our defenses are stripped bare, forced to consider the kind of suffering without meaning that could crush a human spirit. Because in the end, we face disasters like everything else in our lives—searching for meaning. What is left when we are denied a scapegoat or the specter of divine retribution? Our cries of "Why now?" or "Why us?" may never be satisfactorily answered. But if we can look beyond meaning, we'll find a question with profound moral implications: How, in the face of catastrophe, do we help ourselves and the people around us survive and make a better life?

CHAPTER ONE

BRIMSTONE AND FIRE FROM OUT OF HEAVEN

Pompeii, Roman Empire, AD 79

The earth rocked and shook, the bases of the mountains
trembled and reeled because of God's anger.

—Psalm 18

We all know the story of Pompeii. An eruption of poisonous gases and heavy ash covered the Roman city some two thousand years ago, burying people in their houses, completely wiping out the city in a matter of days. We look back and see the inevitability of the destruction and pity the inhabitants for not knowing better. *Who would build a city on the side of an active volcano?* Tourists today visit what might be considered a parable for what happens when you build a community without regard for the threats around you, a place preserved for our edification and amusement. We assure ourselves we wouldn't make the same mistake.

Mount Vesuvius is a classic conical volcano rising over four thousand feet above the Bay of Naples. Its shape tells geologists much of what is going on inside. The massive cone demonstrates that lava comes out faster than erosion can wash it away, so it is active now, and future eruptions are a certainty on the scale of geologic time. To rise up and form a mountain as it has, and not just flow as a liquid over the landscape, the lava must be fairly sticky (or *viscous,* to use the technical term). The sticky lava can hold in gases, at least for a while. That means that eruptions can be explosive. Alternating layers of volcanic ash, the result of explosive eruptions, and cooled lava are needed to grow the tallest mountains—a type called stratovolcanoes.

So why build a city here, where the danger is so great? For the same reason that Seattle lies in the shadow of Mount Rainier, Tokyo looks up to Mount Fuji, and Jakarta is encircled by five active volcanoes, including Krakatau: when they aren't erupting, volcanoes make great homes. Volcanic soils are porous with good drainage and lots of fresh nutrients, producing fertile crops. Deformation of the rocks around a volcano often creates good natural harbors and defensible valleys. Plate tectonics might guarantee that the next event will happen, but which generation will experience the extreme event is determined by chance. And for most human beings, as for the inhabitants of Pompeii in AD 79: if it hasn't happened to me, it simply hasn't happened.

*

Vesuvius's eruption in the sixth century BC led the Osci tribes of that region, and the Roman conquerors who followed, to declare it the home of the god Vulcan. The periodic steam rising from it was a reminder that Vulcan was the smith of the gods, forging their weapons in a celestial furnace. But the volcanic soil was fertile, holding water and supporting some of the richest agriculture of the Roman Empire, and so civilization flourished. Six hundred years without an eruption had made Vesuvius seem the definition of *safe*.

By the beginning of the first century AD, several towns had been built on the side of the volcano, including Pompeii, Herculaneum, and Misenum. The region had been conquered by Rome in the third century BC, and it had become a flourishing, prosperous community. Excavations have found the remains of a thriving commercial center. Frescoes celebrate the craftsmen who wove and dyed cloth, a major local industry. A sprawling, open-air marketplace has been uncovered, complete with restaurants and snack bars. Tax records show that Pompeii's vineyards were much more productive than those around Rome and their wine was sold across the empire. (The first known product brand based on a pun is from Pompeii, a jar of wine labeled "Vesuvinum.")

Wealthy Romans constructed villas there in order to enjoy the seaside. Large public markets, houses of worship, and government buildings reflect a community living well above simple subsistence. Many of the houses excavated in Pompeii are spacious and elegant. Beds were found carved out of marble. Some houses had their own baths, and public baths served the community with water brought in from the Roman aqueduct system. Situated at the end of the Amalfi Coast, Pompeii, even then, hosted the glitterati.

It is from this culture that we get our word *disaster*—literally, "ill-starred." Romans believed that disasters happened because their fate had been written in the stars. The random nature of disasters, relative to the scale of one human lifetime, creates such a level of fear that all human cultures have come up with some means for ascribing meaning to them. When Shakespeare, in *Julius Caesar,* gives Cassius the line "The fault, dear Brutus, is not in our stars, / But in ourselves," he is speaking against a cultural norm that finds explanation of the unexpected in our fates.

Romans were in the hands not just of destiny but of their capricious gods. Like the Greek before it, Roman mythology portrayed the gods as selfish, careless entities, albeit very powerful ones. Disasters happened to an individual because he got in the way of a spat between these powerful beings. Vulcan, the god of fire, was not physically attractive, but he had been given Venus, the goddess of love, to be his wife. Volcanic eruptions, then, were a sign of his anger when he found out about one of Venus's infidelities.

This may have provided an explanation for volcanic episodes, but it was not a particularly reassuring one. It left the people powerless in the face of petty gods and their tantrums. So they attempted to soothe Vulcan—to reclaim a sense of control—in their annual feast honoring him. Vulcan represented fire both in its beneficial uses, such as metalsmithing, and in its destructive power, such as in volcanoes and wildfires (the more common threat to grain storage in the heat of the summer). So with the Vulcanalia, held every year on August 23, they placated the god, offering bonfires and sacrifices to keep destruction from being visited upon their harvest.

In AD 79, as the Vulcanalia was being celebrated by unwitting residents of Pompeii, Vesuvius was entering the final phase of what would be one of its largest eruptions. Our knowledge of the eruption comes from two sources. One is, of course, the evidence preserved in the city of Pompeii itself, fifteen miles outside Naples. The ash from the eruption buried the city over the course of a few weeks, completely destroying the community. Ninety percent of the residents escaped alive, but they abandoned the region, and the existence of the city was almost forgotten. The site was rediscovered and excavated in the eighteenth century, including the corpses of the residents who did not escape.

The second source is a young Roman scholar, Pliny, called the Younger, who wrote letters that have come down to us describing the death of his uncle, Pliny the Elder, during the eruption. The younger and elder Plinys were part of Rome's minor aristocracy, both holding the rank of *equestrian,* entitled to be a knight in the army, and were originally from the Lake Como region in northern Italy. Pliny the Elder served in the Roman army, primarily in Germany, for the first two decades of his adulthood. He never married, but his widowed sister came to live with him after he left the army, along with her young son. The son was adopted by his uncle and took his name, and thus came to be called Pliny the Younger. Pliny the Elder was famous in Rome both for his writings and for his close relationship with the emperor Vespasian. While in the army, he wrote a history of the German wars, with details like how to use a horse's movements to fight with a javelin more effectively. In his later diplomatic career as a ruler of various provinces, he collected information about the history of the regions and their natural features.

Two years before the eruption, Pliny the Elder published his thirty-seven-volume *Naturalis Historiae,* "On Natural History," often called the first encyclopedia. It represented his observations as he traveled the empire, creating one of the largest literary works to come down to us from Roman times. In the preface, he says that "to be alive is to be watchful," and we see that passion in the breadth of topics he catalogs. From a modern scientist's perspective, he may seem a bit credulous (as in, for example, his description of monstrous races of people with the heads of dogs). But he also shows a scientist's passion for knowledge. He finishes his final volume with the words "Greetings, Nature, mother of all creation, show me your favor in that I alone of Rome's citizens have praised you in all your aspects." He seems to have been obsessive about his work, often choosing writing over sleep.

In AD 77, in addition to the release of his *On Natural History,* Pliny the Elder was appointed by the emperor to be head of the Roman fleet berthed in the Bay of Naples. The Pliny household moved to Misenum, at the mouth of the Bay of Naples. From their villa, they had a commanding view of Mount Vesuvius on the other side of the bay. Pliny the Elder directed fleet operations as he worked on revisions to his *Natural History.* Pliny the Younger was completing his legal training, studying with his uncle, and becoming a prolific chronicler himself.

After the centuries of quiet, the latter part of the first century had seen an increase in earthquakes, with a particularly severe one in AD 62. That earthquake had damaged quite a few houses in Pompeii (even in AD 79, some of them were still being repaired).

Numerous earthquakes were felt and recorded in the next decade, and people began accepting them as a normal part of life. At the Vulcanalia celebrations on August 23, AD 79, Pliny the Younger's journal entry noted several earthquakes occurring during the day, but he thought nothing of them, "as earthquakes are common in [the region of] Campagnia." We now know that magma must move from the magma chamber, often several miles deep in the earth, to

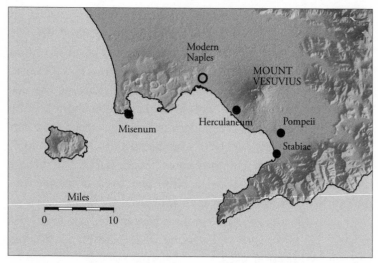

Map of the region of the Bay of Naples, showing Misenum, where the Pliny household lived, and cities all or partially destroyed in the eruption of Mount Vesuvius in AD 79

the surface for an eruption to occur. That movement can be marked by earthquakes, the bulging of the earth's surface, and gas emissions. It can take months, years, or even decades before enough pressure has been built to cause an eruption. (This makes volcanic eruptions more likely to be predicted than many other geological phenomena.)

The next day, August 24, the lives of everyone in Campagnia were turned on end. A little after noon, Vesuvius exploded violently, sending a column of gas and ash high into the sky. Both Plinys observed this from across the Bay of Naples. The Younger wrote, "I cannot give you a more exact description of its appearance than by comparing it to a pine tree; for it shot up to a great height in the form of a tall trunk, which spread out at the top as though into branches."

True to form, Pliny the Elder wanted to see the eruption more closely. He started arranging for vessels of the fleet to help with evacuations and to take him across the bay to make more detailed observations. Pliny the Younger wisely chose to stay at the villa and continue with his schoolwork. As preparations were under way, the Elder received a message from a noblewoman friend whose villa sat in Stabiae at the foot of Vesuvius, begging him to help them escape. He dispatched the galleys to Herculaneum, but he himself took a "fast-sailing cutter." As they approached Herculaneum, cinders and ash fell so heavily that the ship's pilot advised returning to Misenum. Pliny replied that "fortune favors the brave," ordering the pilot to sail for Stabiae, where his friend lived. Winds whipped up by the eruption brought the cutter into port but then made it impossible to leave.

Pliny's friend and her household were terrified by the eruption and the inability of the ship to navigate the rough seas roiled by the turmoil of the eruption. In her villa, Pliny tried to reassure his friends by feasting, bathing, and sleeping while waiting for the winds to abate. But as the eruption grew worse, it became clear that the winds were not dying down. (They were in fact being generated by the eruption itself, although Pliny obviously did not know that.) They decided to try again to get the ship to sea. They ventured back to the shore with pillows tied to their heads to protect them from the falling volcanic ash and molten rock. The sea was still too rough to board the ships, and the air was so foul as to make breathing difficult. Pliny the Elder was overcome and fell to the ground, unable to rise. His friends finally abandoned him and boarded the ship. They were able to escape and thus give the tale to Pliny the Younger. The friends returned three days later and found the body of Pliny the Elder under ash but with no obvious injury. Most scholars have decided he died from a heart attack, perhaps triggered by noxious gases.

*

The type of explosive eruption that throws lava high into the atmosphere, where it solidifies into a variety of particles called (depending on size) volcanic hairs, ash, and bombs, is characteristic of stratovolcanoes. They are found in places where one tectonic plate is being pushed under another, called subduction zones. In the case of Vesuvius, the African continent is slowly moving toward Europe, pushing up mountains from the Alps to the Pyrenees and the Apennines, and pushing the Mediterranean seafloor *under* Italy. As the seafloor is pushed under the continent, friction heats the seafloor, melting it and the sediment carried upon it.

This sediment is a key to understanding such volcanoes. First, compared to lava coming from deeper in the earth and found in other types of volcanoes, stratovolcano lava is rich in quartz, which is a light mineral. So as rocks move around (on geologic time, rocks move a lot), it tends to travel upward relative to the heavier minerals around it. It gradually gets concentrated in the continents (rather than deeper in the earth), and in the sediments that erode off continents. This quartz creates magma with a higher viscosity than the kind found in other volcanoes. And second, the sediments have a lot of water incorporated in them, which in turn becomes incorporated in the magma they create.

The stickier quartz means the lava tends to adhere to itself instead of flowing forth, like we see in pictures of Hawaiian volcanoes. The water means there are more gases and vapor in the lava. Those expand when heated, causing explosions. Krakatau, Mount St. Helens, and Vesuvius are all found in subduction zones, and they all have the potential for these explosive eruptions.

Volcanologists have studied the deposits around Pompeii and the record left by Pliny the Younger and have concluded that the eruption had two main phases. The first was the eruptive column on August 24, a type of eruption now called Plinian. It towered into the air with a massive explosive force but then dispersed sideways and down, pulled by gravity—the pine tree shape that Pliny the Younger noted. From across the Bay of Naples, he said that after the first upward explosion, the ash settled back toward the earth and the day took on "a darkness that was not like a moonless or cloudy night, but more like the black of closed and unlighted rooms. You could hear women lamenting, children crying, men shouting.

Some were calling for parents, others for children or spouses; they could only recognize them by their voices."

Most of the approximately eleven thousand inhabitants of the region left on foot through the darkness, escaping with their lives. When the word came to Pliny the Younger about his uncle's fate, he took his mother (aged and corpulent like her brother) and struggled on foot to get away. Other refugees clogged the roads, floundering through the darkness. Pliny the Younger described a people who believed the end of the world was at hand.

> Many besought the aid of the gods, but still more imagined there were no gods left, and that the universe was plunged into eternal darkness for evermore. There were people, too, who added to the real perils by inventing fictitious dangers: some reported that part of Misenum had collapsed or another part was on fire, and though their tales were false they found others to believe them. . . . I could boast that not a groan or cry of fear escaped me in these perils, but I admit that I derived some poor consolation in my mortal lot from the belief that the whole world was dying with me and I with it.

After several days, Pliny the Younger and his mother escaped to safety, ultimately returning to Rome. Unlike them, some residents decided to stay, at least through the first night. The ash had been falling for a day by that point, and a home protected you from falling rocks. It's easy to see how the decision to stay in your house could seem like a sensible one. What residents of Pompeii and Herculaneum could not know was that the night would bring the second phase of the eruption.

When stratovolcanoes explode, the ejecta usually travel high into the atmosphere, many tens of thousands of feet. As the eruption proceeds, the material gets heavier, and instead of a mushroom cloud high into the atmosphere, the hot gas and ash start to flow rapidly down the mountain (because they are heavier than air, the gas too can flow *down*.) These are called pyroclastic flows, from the Greek words *pyro* meaning "fire" and *clastic* meaning "broken in pieces." The gases move quickly, usually at fifty miles per hour, although three hundred miles per hour has been recorded. They are so hot—around 500°F—that they kill instantly.

A pyroclastic flow is one of the deadliest forms of eruption. It is too fast to outrun, and seems to surprise the victims. The contorted positions of the eighteen hundred corpses buried in Pompeii led early observers to assume the victims had undergone extreme suffering. It is more likely that theirs were instant deaths from extreme heat, followed by their cadavers spasming from heat shock. It was only then that deposits of ash came to bury the corpses in their homes, preserving their tragic story for two millennia.

*

One of the human species' greatest strengths is its ability to theorize. Evolutionary pressure rewarded brains that saw patterns, even in randomness. When we heard a rustle in the grass, we could imagine it was a random breeze and ignore it, or we could hypothesize that it hid a waiting predator and try to escape it. For the many times it was a breeze, the wrong answer made us unnecessarily anxious, but it did not interfere with our survival. For the rare time that it was a predator, the anxious survived, and those who believed it to be random made a fatal error. At a primal level, we abhor randomness because it leaves us vulnerable.

The need to find meaning in randomness extends beyond existential threats. Stars in the sky are spatially random. That you see one star alone in one part of the sky and several stars in a row in another is the simple product of random distribution; it tells you nothing about whether you will see a star elsewhere. Randomness means you cannot use what came before to predict what will be next. We humans make patterns anyway, creating constellations, and then stories to explain the constellations.

And just as the Greeks and Romans invoked stories of the gods to explain the constellations—Orion's Belt, Cassiopeia—so too, as we've seen, did they ascribe geologic phenomena to the divine. This belief explained an otherwise inexplicable aspect of the natural world, relieving the desire to understand why one generation suffered and another did not. However, as Pompeiians found, no amount of ritual could offer you control. Nothing could. (This was a feeling perhaps familiar to many Romans, whose lives could just as easily be destroyed at the whim of their human rulers.)

As Greco-Roman culture unspooled itself, Jewish culture was developing a different conception of God and His interaction with the human world. The Jews rejected the idea of gods as selfish and petty. They believed in an inherently good, loving God with whom a covenant could be made. But if God was good, then how could we explain the human suffering in earthquakes, floods, and volcanic eruptions? The Jewish answer was that the fault must lie in ourselves. Many ancient cultures have defining flood stories, but in the story of Noah, it is the flood victims, and not their god, who are responsible.

The story of Sodom and Gomorrah makes the connection even more explicit. Describing what sounds distinctly like a pyroclastic flow, the book of Genesis reads, "Then the Lord rained on Sodom and Gomorrah sulfur and fire from the Lord out of heaven." It happens because ten good men could not be found among all the inhabitants of the cities. The Bible repeatedly cites earthquakes and high winds as a sign of God's displeasure with man. "The earth rocked and shook," says the book of Psalms; "the bases of the mountains trembled and reeled because of God's anger."

Contemporary Christian and Jewish writers attributed the destruction of Pompeii to the sacking of Jerusalem by Rome nine years earlier. The Roman general who had led the siege and destruction of Jerusalem, Titus, had become emperor exactly two months before the eruption. (A first-century graffito on the wall of the ruins of Pompeii illustrates the connection: "Sodom and Gomorrah," it reads.) Such belief not only justifies why a good God could allow evil to happen; it provides an illusion of control. If disasters are a punishment for sins, then a pure life offers salvation.

Jews and Christians throughout the ages remained mostly satisfied with this interpretation. It cohered with ideas of predestination, a more deterministic worldview. However, as Western theology developed, some found it hard to accept that *no* innocents were killed in natural disasters. Although the apparently pious priest might hide some horrible sin, the babes in arms simply could not.

The ideas of St. Augustine of Hippo, furthered by St. Thomas Aquinas, developed as a means to reconcile ourselves with this dilemma. They invoked God's need to give us free will. God made us free to choose good or evil; He cannot absolve us of our evil after we've made such a choice. We have to live with the consequences of our decisions.

This is straightforward enough as an explanation of war, for instance; but it stumbles a bit when applied to natural disasters. Especially without an understanding of the physical causes, the occurrence of an earthquake after centuries of quiet strikes us as deeply unfair. St. Augustine called such disasters "natural evils" and believed that Creation itself had been corrupted by the Fall of Adam and Eve, so that natural disasters were reflecting the evil choices of the fallen angels. St. Thomas argued that the evil of suffering in disasters was necessary for some goodness, such as bravery and compassion, to be realized. It is for this reason that God allows "natural evils" to persist.

What such arguments did not yet comprehend is that natural hazards are part and parcel of the systems that make life possible on our planet. The distribution of heat through the atmosphere that coalesces into storms is the same movement necessary to bring water out of the oceans, to rain upon land. A planet without earthquakes would be one without mountains and valleys with which to trap clouds, or faults to trap groundwater and bring it to the surface in springs. Natural disasters, as we've seen, are a consequence of the inevitable fluctuations in the natural environment necessary to support life.

An argument for free will makes a different kind of sense today. Suffering from natural disasters *can* be seen as a consequence of human choice. But it is visited upon us for our failure to build sufficiently strong buildings, to adequately maintain our water pipes—at a time when science and experience allow us to know better. In this light, we see the straightforward moral failing of putting personal short-term gain over the health and safety of our family and others in our community.

But for as long as disasters were seen as divinely ordained, even the study of the physical causes of disasters was limited. The belief that earthquakes were sent by God remained unchallenged for millennia, until the evidence to the contrary became too hard for many to refute.

CHAPTER TWO

BURY THE DEAD AND FEED THE LIVING

Lisbon, Portugal, 1755

But how conceive a God supremely good,
Who heaps his favours on the sons he loves,
Yet scatters evil with as large a hand?

—Voltaire, "Poem on the Lisbon Disaster"

In 1755, Lisbon was the fourth-largest city in Europe after London, Paris, and Vienna. Built on the mouth of the Tagus, the port of Lisbon was one of Europe's largest, bringing in riches from the New World. Its throne was fed by gold and diamonds from the mines in its colony in Brazil. After a muddled royal succession, entwined with the Spanish throne that had threatened its independence, Portugal had reestablished its sovereignty and was ruled by King Joseph I. The country was devoutly Catholic, and both the legal and educational systems reinforced the devotion. The universities and most other educational institutions were run by the Jesuits. The Inquisition was still under way, with "autos-da-fé" (acts of faith) that made public spectacles of the penance of heretics, including those whose sins required burning at the stake.

Meanwhile, the Enlightenment was bubbling up through other parts of Europe. With the scientific revolution, a developing intellectualism had arisen, embracing economics, philosophy, political science, and natural philosophy, coalescing into ideas and disciplines that would change the world. From Descartes's mathematics to Adam Smith's economic theories, people were debating the nature of society and trying to improve it. In Portugal, the devout Catholicism and the strong grip of the Jesuits on the educational system set it apart from some of the other great countries, constraining its intellectual evolution.

King Joseph I, who five years earlier had ascended the throne at the age of thirty-six, enjoyed absolute rule. He had been married at the age of fifteen to the daughter of the king of Spain (as his older sister was married to his wife's brother, the future King Ferdinand VI of Spain). King Joseph was by all accounts quite an intelligent man, but he put much of his energy into music and the hunt, the great interests of his wife (which, as we will see, may have saved his life). With little interest in government, the king had appointed three secretaries of state—for internal, foreign, and military affairs—and trusted them to handle most of the decisions of governance. Sebastião José de Carvalho e Melo was the secretary of state for foreign affairs and quickly rose to be the dominant force in government.

De Carvalho had a very late start to public life. The son of a country squire, he was supposed to become a lawyer, studying at the University of Coimbra. He left in frustration, although it is hard to tell whether it was the rigidity of the Jesuit curriculum or the rigors of academic discipline that drove him out. He tried the army, enlisting as a private, but military life also did not agree with him. His service was brief. Lacking direction, he spent a decade whiling away his time as a man-about-town in Lisbon. The nineteenth-century Oxford historian Morse Stephens wrote of him, "His handsome face, great bodily strength, and proficiency in athletic exercises, made him popular in all circles of society in the capital, in spite of his comparative poverty."

He was a popular guest at all the parties of the aristocracy. At least, until he eloped with the niece of one of the most powerful noblemen of the kingdom. Her family tried to get the marriage annulled, but the lady herself was not interested in losing this husband. Her family finally decided to make the best of the situation and arranged for de Carvalho to be made ambassador to London at the age of forty in 1739.

In London, de Carvalho seems to have finally come into his own. He successfully positioned Portugal in the Court of St. James's and nurtured a close relationship with England in spite of their countries' religious differences. His stay in London expanded his view of the world. He saw the commercial might of the British Empire and studied the politics and economics that fostered British success. In 1745, he was recalled to Lisbon and then sent as a special envoy to Austria to negotiate a difficult settlement between the Austrian court and the pope. A French diplomat at the time described de Carvalho's service. "In these affairs," he wrote, "he afforded many proofs of his skill, wisdom, uprightness, and amiability; and especially of his great patience. . . . He is noble without ostentation; wise and prudent. . . . He is a good citizen of the world."

By the time of his return to Portugal in 1749, de Carvalho had become convinced of the value of strong governmental investment in infrastructure in support of the economy—not to mention the value of secularization, especially in education.

Empowered by King Joseph, de Carvalho set about trying to reform the Portuguese state. He established a central bank and worked to foster and protect various industries. He suspected the stranglehold by the Jesuits on Portuguese intellectual thought was holding Portugal back, but it was difficult to oppose them in such a devoutly Catholic country. More than two hundred thousand of Portugal's three million citizens were in a convent or monastery. The Inquisition had shaped Portugal, where blood purity laws kept track of the "New Christians," descended from the Jews and Moors who had been forcibly converted in the fifteenth century, so that "Old Christians" could avoid contaminating their family lines in such marriages.

Early in his tenure as secretary of state, de Carvalho worked with Jesuits to solidify his standing. But he also worked to limit their authority. In 1751, he secured the pope's agreement that Jesuit Inquisitors could not unilaterally impose a death sentence but must get the agreement of the government. By the arrival of All Saints' Day, November 1, 1755, he was already seen as the de facto ruler of Portugal.

*

Many are surprised to learn that a great earthquake, one above magnitude 8, ever took place in Europe. Wherever there are earthquakes, the ratio of large to small earthquakes is stable. That means that places where smaller earthquakes happen are those more likely to see big ones. And while the region surrounding Portugal shows some earthquake activity, there's not nearly as much as in neighboring regions, especially in southeastern Europe.

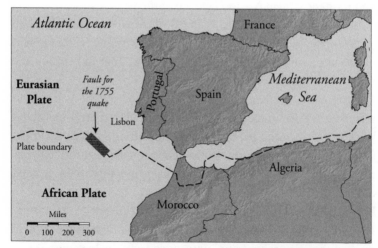

Map of southwestern Europe, showing the plate tectonic boundaries and the probable fault for the 1755 Lisbon earthquake

But in the same way that a high concentration of small earthquakes doesn't necessarily mean a big one is coming soon, the absence of small earthquakes doesn't mean an area is immune to one. How big is limited only by the length of the fault available to break. From our understanding of plate tectonics, we know that Africa is moving into Europe, pushing up the Alps, causing the earthquakes in Greece, Italy, and Turkey, as well as creating the volcanoes of Etna and Vesuvius. West of Gibraltar, this compression continues along the Azores-Gibraltar seismic zone. And indeed numerous earthquakes of magnitude 6 to 7 have been recorded there since 1900. But because they are far enough offshore, those smaller earthquakes haven't caused damage and have been largely overlooked.

The All Saints' Day earthquake of 1755 is by far the largest earthquake that zone has produced in human history. The usual method of determining magnitude today—measuring ground motion on a seismograph—obviously can't help us capture the magnitude of earthquakes that predate such technology. But we know that the energy released in an earthquake is a product of both the area of a fault that moves (more on this later) and also the distance one side moves across the other, the *slip*. We can use that geologic information to estimate the magnitude after the fact. That becomes challenging when the fault we want to look at is underwater, as it was with Lisbon. If it generates a tsunami, we can estimate the size of the fault, and thus the magnitude, from the amount of water in the tsunami. But clearly with old data, any estimates are going to be rough. A third way to determine magnitude is to look at how large an area was damaged and where shaking was felt.

All these approaches have been applied to the Lisbon earthquake, and the smallest estimated magnitude is 8.5. The largest is 9.0. Lisbon was at the northern end of the heavily damaged region, but all the Portuguese coast south of Lisbon was just as heavily shaken. It adds up to a truly massive earthquake, one that has to have had an originating fault on the order of two hundred or more miles long.

*

All Saints' is a holy day of obligation, meaning all practicing Catholics are required to go to Mass. All the churches—and there were many in Lisbon—were running multiple services through the morning to accommodate worshippers. The servants probably went to earlier services so they could be back at work, preparing the feasts to celebrate the day. The gentry and nobility would likely have been at the nine a.m. service. A notable exception was the royal family— their preference for horses and the hunt led them to go to the early service so they could head to their country estate to enjoy the holiday. Those in the churches would have been squeezed into crowded pews, making a quick escape impossible. And, to further complicate matters, the large churches had multiple side chapels, every one of them with their own altar, every altar supporting lit candles.

The shaking began at 9:40 a.m. With such a large earthquake, the rupture had to move down a long fault, which means the shaking likely lasted for three to five minutes. It began with a light level of shaking that grew in intensity. The Reverend Charles Davy, an English clergyman who was living in Lisbon, later described it:

> [The] table I was writing on began to tremble with a gentle motion, which rather surprised me, as I could not perceive a breath of wind stirring. Whilst I was reflecting with myself what this could be owing to, but without having the least apprehension of the real cause, the whole house began to shake from the very foundation, which at first I imputed to the rattling of several coaches in the main street, which usually passed that way, at this time, from Belém to the palace; but on hearkening more attentively, I was soon undeceived, as I found it was owing to a strange frightful kind of noise underground, resembling the hollow distant rumbling of thunder. All this passed in less than a minute.

The sudden slip of a fault in an earthquake twists the ground around it, creating two principal types of waves. P-waves compress the ground, and they travel at the speed of sound. (Sound is a compressive wave itself.) Davy's "noise underground" would have been the P-wave. S-waves, on the other hand, twist the ground. They travel at a slower speed but are bigger than P-waves. The time between the two grows at a rate of one second per five miles of distance the waves have traveled. If the time between the arrival of the P- and S-waves is thirty seconds, we can deduce that the earthquake originated 150 miles away.

The S-wave was a completely different experience:

> I was . . . stunned with a most horrid crash, as if every edifice in the city had tumbled down at once. The house I was in shook with such violence, that the upper stories immediately fell; and though my apartment (which was the first floor) did not then share the same fate, yet everything was thrown out of its place in such a manner that it was with no small difficulty I kept my feet, and expected nothing less than to be soon crushed to death, as the walls continued rocking to and fro in the frightfulest manner, opening in several places; large stones falling down on every side from the cracks, and the ends of most of the rafters starting out from the roof.

Imagine those who sat in pews at church, feeling that first gentle wave. At first, they might have wondered if they'd imagined it. Then a stronger wave came and they looked at their neighbors, trying to decide what to do. No way to get out with all those people around them. Then the S-wave hit and a part of the building collapsed. Lit candles struck people, tapestries, books, the shaking not stopping until the whole building finally collapsed. We can only imagine their terror.

Many people died, crushed by buildings, especially in the churches. A very large aftershock came minutes later to finish the job, leveling many of the buildings that had survived the first round. Those who could had by then exited the buildings, many choosing to congregate near the river quay to be away from the demolition that was taking place around them. It seemed safer, but it put them in the path of a tsunami, which arrived just after the first large aftershock.

Tsunamis begin when the shape of the seafloor changes suddenly. The Lisbon earthquake was probably on a *thrust fault,* where the ground on one side is pushed up and over the other side, creating a new, higher ridge on the seafloor. The water above the ridge is pushed upward, but, being water, it immediately flows down to the lower side. This creates a wave that then moves toward shore. And because the whole column of seawater above the fault is moved, there is a lot of energy in such a wave—the deeper the sea at that point, the more water in the wave. As it approaches shore and the depth of the sea decreases, the height of the wave increases. When it comes to the mouth of a river, as it did with the Tagus in Lisbon, the wave will travel *up* the river, causing sloshing between the banks, with very strong currents that can continue for hours.

Davy had joined many others in heading to the riverside to get away from the falling buildings and was astonished by what he saw:

[T]urning my eyes towards the river, which in that place is nearly four miles broad, I could perceive it heaving and swelling in the most unaccountable manner, as no wind was stirring. In an instant there appeared, at some small distance, a large body of water, rising as it were like a mountain. It came on foaming and roaring, and rushed towards the shore with such impetuosity, that we all immediately ran for our lives as fast as possible; many were actually swept away, and the rest above their waist in water at a good distance from the banks. For my own part I had the narrowest escape, and should certainly have been lost, had I not grasped a large beam that lay on the ground, till the water returned to its channel, which it did almost at the same instant, with equal rapidity.

When it seemed like it couldn't get any worse, it did. All those candles, lit on the altars to celebrate the Mass, had begun setting flame to carved wooden statues, embroidered altar clothes, old prayer books. The flames spread, and there was no organized approach to fighting them. By nightfall, the fires had engulfed all that remained of the city and continued to burn for six days. Eighty-five percent of the buildings in Lisbon were destroyed either by the earthquake or by the fires. The damage was worse along the river, where the loose river sediments had amplified the shaking. As this was the historic center of the city, these were the most significant buildings—the palaces, archives, and churches.

Nor was Lisbon the only city affected. Most of the towns on the southern coast were badly damaged or destroyed. As is common with historic events, estimates of casualties vary widely, but the soberest sources suggest forty to fifty thousand dead, with three-quarters of those deaths occurring in Lisbon.

*

While the Lisbon earthquake is remembered as the greatest natural disaster to strike Europe, it should also be remembered as the first significant response of a central government to a natural disaster. It still stands as one of the most effective. The royal palace in Lisbon was completely destroyed, but as the royal family had left town earlier that morning, the court eventually gathered around the king at the small palace of Belém, outside Lisbon. The king is said to have exclaimed to de Carvalho, "What is to be done to meet this infliction of Divine Justice?" De Carvalho's calm reply became legend. "Sire, we bury the dead and feed the living." He quickly took charge of what remained of the government, and by all accounts, his decisive action inspired only gratitude and obedience in that traumatic time.

For the next eight days, de Carvalho lived in his carriage, mobilizing a response and reestablishing control. Guards were stationed around the outskirts of Lisbon to prevent able-bodied residents from leaving, forcing them instead to stay in the city and help: removing debris and creating adequate shelters for the survivors. To prevent looting, gallows were constructed at several high points of the city, providing summary justice to those caught in the act. More than thirty people were executed in the next month.

De Carvalho promulgated two hundred decrees to maintain order and start recovery. Some were done so quickly that they were written against his knees in pencil and sent to their destinations without even being copied. These decrees covered many of the functions familiar to modern responders—arrange shelter and food for the homeless, treat the wounded, forbid exploitative pricing, reestablish schools and churches. There were too many bodies to bury before decomposition turned them into a public health nightmare, so the bodies were cast into the sea tied to heavy weights, despite Jesuit opposition.

De Carvalho's greatest insight was recognizing the need to expedite recovery—to give people hope and, just as important, to keep them in place. On December 4, 1755, barely a month after the earthquake, the chief engineer of the realm presented to the king four options for the recovery. They included abandoning Lisbon; rebuilding with recycled materials; rebuilding while widening some streets and making improvements to reduce future fire damage; and building a new city completely. With aid from other European countries, especially England, the king chose the most ambitious plan. In less than a year, Lisbon had been cleared of debris and reconstruction had begun.

De Carvalho became the hero of Lisbon. The king named him prime minister to give him complete power to enact the reconstruction. A few years later, he was given the title Marquis de Pombal. He succeeded in the complete rebuilding of Lisbon and was widely credited with saving many lives in the aftermath. He required that new buildings be constructed to withstand future earthquakes. Scale models were built, their resistance to shaking tested by having the army march around them. (The style of construction that resulted is called Pombaline in his honor.)

For his part, King Joseph remained psychologically scarred by the earthquake. He developed severe claustrophobia, insisting on living the rest of his life in tents. He relied ever more on de Carvalho to run the country and never questioned his work. Only after the king's death did his daughter, Queen Maria I, begin the construction of a new royal palace.

De Carvalho also undertook the first scientific survey of an earthquake. He sent queries to every parish church with a set of questions: When did the earthquake start and how long did it last? How many people died? Did the sea rise first or fall? The data from his survey has been used by modern scientists to better estimate the source and magnitude of the earthquake.

De Carvalho's success cemented his political authority and enabled him to enact his ambitious plans for the modernization of Portugal. His biggest coup was removing the Jesuits from their positions of power. He deemed them a danger to an absolute monarch but also an obstacle to the country's intellectual advancement. Before the earthquake, he had succeeded only in establishing that the government had to approve an execution ordered by the Inquisition. Just two years after the earthquake, de Carvalho had the Jesuits barred from the Portuguese court altogether. The next year they were barred from engaging in trade, and in 1759, having caught them engaging in a plot against the king, he succeeded in confiscating their holdings and secularizing all the universities.

*

The Lisbon earthquake's physical shaking was felt across much of Europe, as far north as Scandinavia. Occurring at the height of the Great Enlightenment, the earthquake brought about just as profound a philosophical shaking, resonating across broader Europe for decades. At a time when the roles of rational thought and religious faith were being questioned, the deep feeling of unfairness generated by the Lisbon earthquake had a marked impact on philosophy and science. Many historians suggest the earthquake caused a fundamental shift in Christian thought. The political theorist Judith Shklar wrote, "From that day onward, the responsibility for our suffering rested entirely with us and on an uncaring natural environment, where it has remained." The moral philosopher Susan Neiman called it "the beginning of a modern distinction between natural and moral evil."

The Lisbon earthquake was not, however, a universally secularizing force. The causes and consequences of the earthquake were very much in the eyes of the beholders. The French philosopher Voltaire, for one, was deeply moved by the human suffering in the earthquake and quickly wrote his "Poem on the Lisbon Disaster," published only weeks later in December 1755. He rejected the idea that a benevolent God could cause the suffering seen in Lisbon:

> *What crime, what sin, had those young hearts conceived*
> *That lie, bleeding and torn, on mother's breast?*
> *Did fallen Lisbon deeper drink of vice*
> *Than London, Paris, or sunlit Madrid?*

He rejected the prevalent ideas of philosophical optimism—that God created a good world, that any apparent evil was deliberate, that, as Alexander Pope wrote, "What is, is good."

> *And o'er this ghastly chaos you would say*
> *The ills of each make up the good of all!*
> *What blessedness! And as, with quaking voice,*
> *Mortal and pitiful, ye cry, "All's well!"*

Voltaire's rejection of God's guiding hand here is often interpreted as advocacy of atheism. Indeed, the view of disaster as divine punishment is so deeply ingrained in the Western psyche that to reject it is seen as equivalent to rejecting the very concept of

God. But Voltaire was a theist (though admittedly one who decried much of the work of organized religion).

Voltaire's writings inspired responses from the philosophers of Europe, including letters from Rousseau. Rousseau adopted a more traditional view of God's involvement—essentially that there was purpose to the suffering, or else that the suffering simply wasn't that severe. Rousseau was already espousing naturalism—that much of the ill in human life came from being settled in cities and divorced from the fundamental peace of nature. In keeping with this idea, he argued that much of the suffering was caused by the decision to build tall houses in close proximity, and thus was a result of nothing so much as our free will. Further ill-advised human choices are cited by Rousseau—decisions to return to burning buildings to retrieve valuables, for instance. He completes his discussion by arguing that we cannot reject God's influence or his goodness because we do not know what worse suffering the victims were spared by not living past that day. (One has to wonder how depraved Rousseau thought urban life to be that he could imagine the Lisbon earthquake as a lesser evil.)

Whatever its impact on philosophy, the Lisbon earthquake furthered the notion that the physical world could be described and understood through the scientific method. It was felt by budding scientists across Europe and led to a flowering of hypotheses about its possible physical causes. Most drew on Aristotelian ideas about vapors in the earth, as the winds were one of the fastest-moving features of nature that could be observed. (It was not until the San Francisco earthquake in 1906, when a fault could be clearly seen on land, that the role of faults in creating earthquakes was first proposed.) Still, the Lisbon disaster led to significant advances, including the first earthquake catalogs and the first recognition that earthquakes were not spatially random—that areas prone to earthquakes could be recognized.

Outside intellectual circles, however, evidence suggests that the larger sphere of humanity continued to respond to random threats with familiar ideas of divine retribution. The timing of the earth-quake, on a holy day of obligation, at the exact time of the morning when churches would be at their fullest, seemed too significant to be ascribed to simple chance. It invited the question: Why should the pious in their churches be stricken while the prostitutes in the nearby red-light district were, at least relatively speaking, spared?

Modern science offers an explanation for even this discrepancy. In fact, we have three. First, Lisbon was originally settled as a seaport, and as noted, its first buildings were constructed in the sediment along the riverbank. Loose river sediment passes seismic waves more slowly than stiff soil or hard rock. To carry the same amount of energy at a slower speed, the waves have to get bigger. We see amplification by a factor of ten or more in very loose soils. These soils are also subject to liquefaction. When soil is shaken, the grains settle closer to each other. (We observe the same phenomenon when we fit a whole bag of flour into a slightly too small canister by tapping it on the kitchen counter.) If the soil was saturated with water, as it would have been near a river, the compression caused by that settling would squeeze out the water held between the soil grains. That would raise the pressure in the water, and, if the pressure got high enough, the sand could temporarily become quicksand, flowing like a fluid until the water could get out of the compressed space. Quicksand does a very bad job of holding up buildings. Reverend Davy, who described the tsunami, also spoke of *sand blows*—water and sand shooting into the air—which are a characteristic feature of liquefaction.

Second, very big earthquakes produce more low-frequency energy, causing more damage to very large buildings than to small ones. Third, the churches were made of stone, while the brothels in the city's red-light district were more likely to have been built of wood and were thus more flexible and able to withstand shaking.

Of course, people looking for an explanation of God's actions in 1755 didn't have the benefit of these ideas. They were living in a European society divided between the Protestants, who saw the Catholics as papist idolaters, and the Catholics, who had turned to the Inquisition to preserve the true faith against the ravages of Protestantism. In Catholic Portugal, many took the earthquake as a sign that they had been insufficiently zealous. Father Gabriel Malagrida, a Jesuit and a leading preacher of the day, argued that God had destroyed Lisbon because they had allowed too many Protestants into the city. It is significant that of the thirty-four people executed during the chaotic aftermath of the earthquake, most were Protestants. Voltaire ridiculed this in *Candide:*

> After the earthquake had destroyed three-fourths of Lisbon, the sages of that country could think of no means more effectual to prevent utter ruin than to give the people a beautiful auto-da-fé; for it had been decided by the University of Coimbra, that the burning of a few people alive by a slow fire, and with great ceremony, is an infallible secret to hinder the earth from quaking.

In the Protestant world, a response to the earthquake was much easier to offer. Here was God's proof that Catholics really were idolaters, the Inquisition inspired by the Devil. John Wesley, a noted English preacher and founder of Methodism, opined that there was nothing that brought sinners to greater attention to God than an earthquake. About Lisbon, he wrote:

> And what shall we say of the late accounts from Portugal? That some thousand houses, and many thousand persons, are no more! that a fair city is now in ruinous heaps! Is there indeed a God that judges the world? And is he now making inquisition for blood? If so, it is not surprising, he should begin there, where so much blood has been poured on the ground like water! where so many brave men have been

murdered, in the most base and cowardly as well as barba-
rous manner, almost every day, as well as every night, while
none regarded or laid it to heart.

A belief in God's retribution had very real consequences for
the people of Portugal. Its conflicts with Spain had led Portugal
to develop strong bonds with some of the Protestant countries of
Europe. The ambassadors to Portugal from Great Britain and The
Hague were the first to have audiences with King Joseph after the
earthquake, and, moved by the great suffering they had seen, they
wrote appeals to their home countries for aid to the Lisboans. King
George II of Great Britain pledged the huge sum of £100,000 as
immediate aid. But for their part the Protestant Dutch government
declined to help. By their Calvinist thinking, if God had chosen to
punish Portugal for its Romish idolatry, they had no place interfer-
ing. God had rightly decided what level of suffering they should
receive.

CHAPTER THREE

THE GREATEST CATASTROPHE

Iceland, 1783

I thought it would be unfortunate if these memories should be
lost and forgotten upon my departure, as have so many other
works of God which have, for lack of care, been lost forever.

—Jon Steingrimsson, Introduction to his autobiography, 1785

Of all the natural disasters, volcanoes have the greatest potential
for massive physical disruption. No single event demonstrates
that better than the Laki eruption in Iceland in 1783–84, which
researchers believe is the deadliest natural disaster in human history. The total death toll was in the millions, and the devastation
spanned the globe. We are lucky, then, that volcanoes occur only in
a few specific locations. Still, how could a volcanic eruption in an
isolated corner of the North Atlantic—an island with a population
of only fifty thousand where eruptions happened on average every
three to five years—have led to so much death and destruction?

To understand the answer, you have to consider the role volcanoes play in our planet's ever-shifting geography, and the plate
tectonic dance. Volcanoes show up in three different plate tectonic
settings. The first lies under the ocean, forming what are known as
mid-ocean ridges. These are places where the big plates are moving
away from each other and hot magma rises from the mantle, deep
in the earth, to fill in the space left behind. The magma in the man-

tle is dense, so the resulting rock (basalt) is heavy. It sinks a bit into the earth's partially molten mantle. Consequently, you find heavier rocks at the earth's lowest elevations, the seafloor, and lighter minerals at higher elevations, the continents. (It also explains why these volcanoes are *mid-ocean* ridges.)

Harry Hess, a geologist at Princeton University and a rear admiral in the Navy Reserve, had one of the great insights of the plate tectonics revolution of the 1960s. He realized that the volcanoes at the mid-ocean ridges were actually creating *new* seafloor, what he called seafloor spreading. Scientists had been speculating about the movement of continents for decades, since the geologist Alfred Wegener proposed the notion of continental drift in 1912. He had recognized the similarity of rocks and fossils in Africa and South America, so he suggested that the continents must somehow be pushing themselves apart, plowing through the oceanic crust, like icebreaker ships through a frozen lake. However, because the continental crust is made of rock that's not just lighter but weaker than the ocean floor, pushing continents through the earth's crust would be like pushing a marshmallow through a brick. Hess's insight was to see that continents were not drivers but simply passengers carried along as the lithospheric plates moved over the mantle. His argument is supported by evidence we find on the ocean floor, which is nowhere more than 200 million years old. (By comparison, the oldest rock on the continents is 3.7 *billion* years old.)

But this raised a new question: If continents were shifting as a result of the formation of new crust, what was happening to all the old crust? The earth isn't growing bigger, accommodating more and more oceanic crust.

The answer lay in *subduction zones.* This is where two tectonic plates collide and one gets forced *down,* into the earth, where it is melted and eventually recycled, at rates of one to a few inches per year. Rocks created at the mid-ocean ridges exist for anywhere from a few million to 200 million years, before being subducted into the earth to be reabsorbed.

It is this motion that's responsible for the second occurrence of volcanoes, those that lie *above* subduction zones. As with Vesuvius, a subduction volcano is one that results from a plate being gradually forced down into the earth, until friction with the overriding plate melts the rock, which then rises through the upper plate, emerging as a volcano. In addition to those in Italy, subduction volcanoes compose the circum-Pacific "Ring of Fire," including the many volcanoes of Japan and the Pacific Northwest.

The island of Iceland is an exception. It represents the third tectonic setting in which we see volcanoes: *hot spots.* There are a few places in the earth's mantle that are particularly, inexplicably hot. Hot things rise, of course, and so plumes of magma come up from deep in the earth, irrespective of what's above them. Hawaii, Yellowstone, Galápagos, Réunion Island, and Iceland are the most notable of these hot spot volcanoes. Iceland is unique in that its plume coincides with a mid-ocean ridge.

The island of Iceland as we know it exists because a hot spot has caused considerably more magma to head to the surface than in the rest of the Mid-Atlantic Ridge, which is still below the ocean's surface. The oldest rock in all of Iceland is 13.5 million years old (younger even than most of the ocean floor). Every bit of the country, which is the size of Tennessee, was created as a result of a volcanic eruption. It has dozens of subpeaks, but it's fair to say that the whole country *is,* essentially, one very big, active volcano.

*

Formed in the middle of the Atlantic, Iceland has never had a landbridge to other continental areas. As the polar ice cap melted at the end of the last ice age, about twelve thousand years ago, Iceland emerged as a pristine land without people. Birds, plants, and sea mammals abounded, but the only land mammal found in Iceland before human settlement was the arctic fox. The first humans to find Iceland were most likely Irish monks. Irish tradition says that St. Brendan the Navigator sailed through the North Atlantic in the sixth century AD and arrived at an island he called Tila, where foul-smelling rocks fell on him and his fellow monks. When the Vikings from Norway arrived in Iceland in the mid-ninth century, they found Irish hermits near a dale on the southeast coast that they called Kirkjubaer, or "church farm."

The first of Iceland's Scandinavian settlers were two men and their households. Ingólfur Arnarson and Hjörleifur Hróðmarsson were blood brothers, having sworn allegiance to each other in the Viking custom that demanded that each avenge the other in the event of one's murder. The story of their settlements is recorded in the *Landnámabók* (Land-taking book), written a few hundred years later. It functions as a foundational myth of the country of Iceland, the stories they tell themselves about who they are.

The book reports that Ingólfur had brought "high-seat pillars," with the story of his family carved into their wood, from his old home. On approaching Iceland, he followed tradition and threw the pillars overboard, swearing to make his settlement where the pillars washed to shore, so that the gods could show him where he should live. Hjörleifur did not indulge in such superstition; he came into the first good harbor.

Ingólfur's pillars washed ashore in a bay bordered by steaming hot springs, which he named the Bay of Smoke, or Reykjavík. Hjörleifur meanwhile settled on the southern shore near the old Irish monks' Kirkjubaer. Struggling to establish a farm, Hjörleifur pushed his Irish slaves so hard that they rebelled, murdering him. Ingólfur discovered what had happened and avenged his blood brother, executing the slaves. He lamented Hjörleifur's sad end but found his death only to be expected for someone who ignored his people's rituals.

The settlement of Iceland was completed over a few decades, from AD 874 to 930, involving about ten thousand Norwegians and

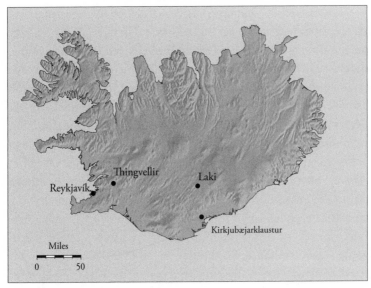

Map of Iceland, and the Laki craters

their Celtic slaves. Norway was in turmoil at the time, with its king, Haraldur the Fairhaired, trying to consolidate his power over the chieftains. Up until then the king was considered more of a "first among equals," but Haraldur conquered several small surrounding kingdoms and demanded taxes from his fellow chieftains. Meanwhile, a burgeoning Norwegian population was running out of easily accessible land. And so a new land with open fields (and no natives) was a great draw, in spite of its distance from Norway—or perhaps because of its great distance from the king.

Because so many of Iceland's first settlers were rebelling against King Haraldur's newly imposed taxes, the resulting culture was remarkably egalitarian for its time, with local chieftains and farmers, but without a king. The chieftains (in old Norse, *goði*—from the word for "god") played both a political and a priestly role. No cities were formed, not even villages. The people were spread out in farming homesteads, gathering together once a year at the Althingi— the world's first parliament.

Since those first "foul-smelling rocks" fell on St. Brendan, volcanic eruptions had been a part of Iceland and its culture, albeit a very dangerous part. Farms found themselves covered in lava; eruptions under nearby glaciers resulted in sudden melts, causing massive surge floods. But the same forces that caused so much disruption also provided an important source of heat that made life there viable. Iceland had been forested when the Vikings first arrived; but the minimal yearly growth of trees, through the island's very short summers, and the tendency of sheep, brought on longboats from Norway, to feed off young saplings meant the country had been almost completely deforested within a few hundred years, leaving Icelanders without wood for fuel. Many homes were deliberately built around the hot springs formed by volcanoes, so the steam would warm the house. Even today, much of Iceland's economy is powered by the essentially unlimited energy derived from geothermal plants. Icelanders call their home the Land of Fire and Ice.

Volcanoes also created a landscape Icelanders called Thingvellir, or "Parliament Plains." A verdant valley, Thingvellir is bounded by vertical basaltic cliffs. (It is, in geological terms, a *graben*: a valley formed by the parallel faults around it.) In this natural amphitheater, the Speaker of the Law (the lawyer who adjudicated disputes brought to the Althingi) could speak to and be heard by the assembled crowd. The Althingi met here each midsummer from 930 to 1798, a proud symbol of Iceland's independence and egalitarianism.

Like the discovery of Reykjavík through Ingólfur's high-seat pillars, the annual meeting of the Althingi in Thingvellir is a story central to the Icelandic identity. It also reflects the Icelanders' more practical side. At the end of the tenth century, Iceland was struggling with conflicts between practitioners of the old Norse religion and the Christian converts, in part because of the role of the *goði* in both politics and religion. The arguments came to a head when, in the summer of 1000, the Althingi debated whether the country should become Christian. During the debates, a rider came in and told the assembly that a volcanic eruption was under way nearby.

Some called out that the land gods were upset by their debate, a clear sign they needed to reject Christianity. Snorri Goði, the Speaker of the Law that year, replied, "What were the gods upset about when the rock we are standing on erupted?" After the laughter died down, the assembly voted in favor of Christianity.*

*

In the centuries that followed, repeated eruptions took lives and damaged livelihoods. Combined with a mini ice age and the arrival of the plague in the fourteenth to sixteenth centuries, Iceland was brought close to starvation, dependent on meager aid from first the Norwegian and then the Danish kings, with whom the Icelanders had realigned. But unlike the Greenland settlers who died out in that period, the Icelanders survived to build a resilient community. By the mid-eighteenth century, Iceland's population had grown to fifty thousand people, still spread out in farmsteads with few villages. The conversion to Christianity had integrated the church into Icelandic life, with churches on farms worked by the pastors and their families.

The old Irish settlement Kirkjubaer, where Hjörleifur had landed and met his fate, had been the site of a convent for a couple of centuries in the Middle Ages. Thus "cloister" was added to its original "church farm" name, producing *Kirkjubæjarklaustur*. It was a rich and prosperous settlement ministered to by a popular pastor, Jon Steingrimsson.

On the morning of Pentecost, June 8, 1783, Pastor Jon was riding his horse to his church, ready to preach on the coming of the Holy Spirit, when he saw a great black cloud rising to the north. Within minutes the darkness enclosed him and ash began to fall.

---

* This story is folklore, recorded in the Icelandic sagas, written down a few centuries later. By the time it was written, Icelanders would have seen many eruptions and the solidification of the lava into rock and known the relationship well. This is perhaps the first recorded instance of Europeans laughing off the notion of divine retribution.

To Jon, God's patience had run its course. The hour of affliction had arrived.

Jon Steingrimsson is a hero in Iceland, his story taught in schools. He exemplifies courage and calm in the face of a disaster that threatened to push Icelanders to extinction. He exhibited at once an Icelandic favor for superstition and the same skepticism seen nearly eight centuries prior in Snorri Goði. He kept detailed journals in which he chronicled his dreams, which he believed to be portents of the future. He saw the Laki eruption as a punishment by God for the sins of the Icelanders. But the journals also recorded detailed descriptions of the eruptions and other volcanic phenomena, making them an important primary resource for modern volcanologists.

The journals should be required reading for anyone interested in disaster response and recovery. Like records of the Lisbon earthquake before it, they demonstrate how a disaster only begins with the natural hazard. During the event itself, damage is inflicted, lives are lost, heroes are needed to rescue victims. But it's only after the event has passed that the more difficult phase begins, the recovery and reconstruction that requires courage, perseverance, and leadership. Jon excelled during both phases, demonstrating what a disaster can demand and what difference an individual can make.

The eruption that began that Whitsunday morning continued for eight long months. It deposited a fifty-foot-thick blanket of lava over six hundred square miles, an area more than half the size of Rhode Island and representing one-sixth of the total area of Iceland. Most of the lava came in the first forty-five days, when it was described as flowing as fast as a river in spring flood. It came from a series of eruptions, from ten different fissures, with each following the same pattern: First a series of earthquakes would occur, spanning a few days or weeks. Then a fissure would be pushed open and lava would begin rising through the water naturally present in the ground. Next, this interaction between lava and water would create an explosive eruption, and as the eruption through each fissure progressed, the groundwater would eventually evaporate, causing the lava to transition into surface flow.

The result was an alternating series of explosive and surface phases. In the eight months that followed, Laki produced more than three times as much lava as Mount Kilauea in Hawaii has in thirty years of continuous production.

Jon described in his journal the earliest days of eruption, as the lava approached, and when it seemed the earth itself was being torn apart:

> This past week, and the two prior to it, more poison fell from the sky than words can describe: ash, volcanic hairs, rain full of sulfur and saltpeter, all of it mixed with sand. The snouts, nostrils, and feet of livestock grazing or walking on the grass turned bright yellow and raw. All water went tepid and light blue in color and gravel hillsides turned gray. All the earth's plants burned, withered and turned gray, one after another, as the fire increased and neared the settlements.

In addition to the devastation, he recorded his ministrations—both spiritual and medical—to his flock. (He had trained himself in what medical arts were known at the time.) Jon rode ceaselessly around his district, checking on his parishioners' welfare. Meanwhile the lava continued to flow, progressing ever closer to Kirkjubæjarklaustur and Jon's church.

On July 20, Jon gathered his congregation, believing it would be the last time they would be able to worship in their church. By that point, the lava had flowed down the river valley closest to the church. The end seemed near. Many of the parishioners had lost their farms, and others were dying from the gases emitted by the lava. Some asked for the church doors to stay open so they could see when the lava would be upon them and escape.

Jon gave a sermon that has since been memorialized as the "Fire Mass" (*Eldmassan*). Jon himself didn't dwell much on the event in his journal and offers only a limited description of what was said. We know that he began by asking all to "pray to God in correct piety, that He in His grace will not want to destroy us in haste." He called on them to remember that however bad things were, God was greater. Their job was to suffer patiently what was laid on them and to trust in God's mercy.

What more he said has been lost to history. But his speech had an outsized impact on his legacy. When the service was over and the parishioners came back outside, they found that the lava had stopped its flow before engulfing his church. (Modern investigations have shown that the lava flowed into a river with enough water that the lava froze into a natural dam before the water in the river had all turned to steam. The dam diverted the rest of the lava and kept it away from the church.) Jon was hailed as a miracle worker, to be known to future generations as the Fire Priest.

The threat had far from passed. Lava continued to flow for six more months. Later phases spanned other river valleys, spreading the devastation through much of the southeast, in what had been the most productive agricultural region. When the lava flows finally abated in early 1784, the feeling of relief was short-lived, as the poisonous gases only continued to wreak havoc. Called *Móðuharðinin*, or the Mist Hardships, they almost destroyed the country. More than 60 percent of the livestock died, including 80 percent of the sheep, the primary source of meat. Ten thousand people, over one-fifth of the nation's population, died from the gases and the resulting famine.

*

Among Laki's emissions, two gases, hydrogen fluoride and sulfur dioxide, were ejected in large quantities. Fluorine is an element that affects the development of teeth and bones and, in small quantities, strengthens them to our benefit. Hydrogen fluoride can break down into fluorine, and it is highly water soluble, which means that it dissolves in the rain and can coat ash particles. (Even today, Icelandic farmers will leave out a bowl of water and, if they find ash in it, presumably from a distant eruption, will bring their animals indoors to protect them from the fluorine.) The resulting fluorine is then carried into water supplies and absorbed into plants.

In Laki, eight million tons of hydrogen fluoride settled over the country. In these massive quantities, fluorine poisons the body, deforms the bones, and destroys the teeth. In his journal, Jon describes the hooves of the animals as they rotted out from under them. Desperate for food, some people ate this contaminated meat, and many of them died as a result. The people on the coast who depended on ocean fishing fared better than those dependent on farming. The fluorine contaminated the pasturage and the fish in freshwater streams, but oceangoing fish remained healthy. For the next two years, things only deteriorated. As a colony of Denmark at the time, Iceland had little local organized government and there was no countrywide effort to get food to people in need.

Pastor Jon did his part to save his congregation. He traveled to Reykjavík to get help for his district and was given some money by the Danish representatives, though much of it had been stolen from him by the time he returned. He continued to minister, visiting farms, creating medicines, documenting the suffering and starvation, and, increasingly often, burying those for whom help was insufficient.

Jon was committed to ensuring that every soul received a Christian burial, even when he had no one to help him. He had the only horse still healthy enough to carry the bodies back to the churchyard, sometimes five or ten in a week, and he kept a record in his journal of every death. His spirit was almost broken when his beloved wife of thirty-one years, Thorunn, became one more statistic in his journal. Alone, without fuel for his lamps, his hands and feet swollen from frost, he wrote of the temptation of suicide.

After two years of this, in the autumn of 1785, when it seemed impossible to survive, Jon organized one last trip to the coastline to see if anything edible could be found. A man and two boys went

ahead to scout the situation. The rest of the group was amazed
when they caught up to find that their scouts had killed so many
seals that they'd need 150 horses to take them home. It was a god-
send, enabling the community to survive the winter and begin the
climb back to normalcy.

Jon's community and the other southeast farmlands suffered the
most, but it was a national crisis. It took over a year for the Danish
government even to send an envoy to see what was going on, and
their aid was minimal. So much farmland was under lava, and so
much more poisoned by gas, that large parts of the population were
forced from their ancestral homes to search for uncontaminated
land. Iceland became a nation of refugees with no place to go.

Here we see the defining element of a catastrophe, when the
extent of the damage is so large that human society itself is
threatened—when half a country loses its farmland and its means
of basic subsistence. The mass migration of people across the country disrupted much of the functioning of the government and
church. In many places, the baptismal and funeral records were
lost (perhaps because there was no one left to record the pass-
ings). Other changes in Icelandic society have been attributed to
the "Mist Hardships." A long history of traditional dancing, called
víkivaki, died out at about this time. A noted Icelandic historian,
Gunnar Karlsson, believes that the nation was in such shock that
they had no heart for it.

The economic and social consequences of a disruption to human society after an event can dwarf the physical destruction of
the event. Crucial to a society's survival is how quickly it can recover from the disruption and restart the regional economy. Iceland's society could have completely collapsed, its population either
displaced or extinguished. It was through the effort of people like
Jon Steingrimsson to redistribute food, to provide medical help and
inspire hope, that the nation survived.

Like de Carvalho in Lisbon, Jon Steingrimsson was critical to
the survival of the society as it recovered from disaster. But this is

not how he is remembered. Our emotional response to heroism in the moment of crisis, amplified by our fear of the random bolt out of the blue, is hard to overcome. His efforts toward the nation's recovery may have had the greater impact, but Jon is remembered first and foremost as the Fire Priest, the man who preached a sermon that could make lava stop in its tracks.

*

In addition to fluorine, Laki produced an abundance of sulfur dioxide. Sulfur dioxide is a heavy compound, with more than twice the density of water. It therefore takes a lot of energy to push it high into the atmosphere. In the less explosive phases of the Laki eruption, the sulfur dioxide rained down on Iceland, burning holes in leaves and furthering the destruction of the crops. But the eruptions were often powerful enough that a large part made it into the stratosphere, carrying it to Europe and beyond.

It was this ejection of gases into the upper atmosphere and stratosphere that unleashed the full extent of Laki's devastation. It caused so much disruption in Europe that 1783 was called *annus mirabilis,* a year of wonders. A haze of sulfur dioxide, sulfates, and ash first appeared over the Faroe Islands, halfway between Iceland and Norway, on June 10, two days after the eruptions began; spread into France by June 14; and moved across all of Europe by the end of the month. It stayed throughout the summer, with newspapers reporting a smoky fog for weeks or even months in some locations. By the fall, many newspapers in England and France were reporting an inexplicable illness felling people across those countries, characterized by burning throats and constricted breathing. So many farm laborers were struck down that farmers struggled to bring in their crops. A comparison of British death records that summer with other years suggests that as many as twenty-three thousand people were killed by the Laki eruption in the UK alone.

Alexandra Witze and Jeff Kanipe, in their book *Island of Fire,* describe the widespread poisoning and disruption of weather this

occasioned in Europe, citing a contemporary source. "Such multitudes are indisposed by fevers in this country, that farmers have
with difficulty carried in their harvest, the labourers having been
almost every day carried out of the field incapable of work, and
many die."

As bad as this sounds, the sulfur that stayed in the stratosphere
caused even more harm. It oxidized there into sulfuric acid and
condensed into sulfate aerosols. In the lower atmosphere, sulfates
are washed out of the atmosphere relatively quickly by rain. But
above the main climate systems, in the much drier stratosphere,
particles could be transported around the world, staying aloft for
years. These sulfate particles are just the right size to scatter incoming sunlight, sending some of it back into space and, consequently,
cooling the ground below. Volcanic eruptions that send a lot of sulfur into the stratosphere can have a substantial impact on the global
temperature. Mount Pinatubo, erupting in 1991, cooled the world
by 1.5°F, with an impact that could still be felt three years later. We
don't have the same kind of precise measurements for Laki, but we
know it erupted six times as much sulfur dioxide as Pinatubo and
got a higher percentage into the stratosphere.

The following winter was extraordinarily cold, leading to more
deaths from exposure and starvation. From London to Vienna,
newspapers reported people found frozen to death in the streets
and houses, buried under snow. Major rivers froze over, halting
transportation, then flooded in the spring. There was political fall-
out when Marie Antoinette, the French queen, was quoted com-
menting on how the snow-covered streets made it easier to move
her sledge—it created such a political uproar that her husband, the
king, was forced to offer large donations to flood victims to quell
the unrest. Europeans fared little better the following summer. Per-
sistent cold and resulting crop failures led to famine in much of the
continent. Famine in France was a significant contributor to the
social disruption that culminated in the French Revolution.

The disruption didn't stop there. The monsoons that bring life-giving rain to much of the tropics are generated by the temperature difference between continents warmed by summer sun and the cool of the oceans. The blocked sunlight led to cooler continents, reducing the energy in the monsoons. In Egypt, the missing monsoons meant the Nile didn't flood as it normally would, causing widespread drought and famine. The nation lost one-sixth of its population of 3.6 million people. Meanwhile, a massive famine in India killed close to 11 million people, and one in Japan killed more than a million people. (Both the Indian and Japanese famines were compounded by what was probably a strong El Niño condition, so we cannot assign blame solely to Laki.)

All told, more than a million people, perhaps many more, died because of the Laki eruption. More than ten thousand people, almost a quarter of the population, died in Iceland, and the majority of the people lost their homes and livelihoods. Perhaps as many as one hundred thousand people died from exposure to poisonous gases. More than one hundred thousand more died from cold, flooding, and hunger. Millions died from famines exacerbated by the eruption. The full extent of the damage can never be known.

*

Volcanoes have the capacity, unique among natural hazards, to cause global impacts because of their ability to affect the composition of the stratosphere. All hazards affect the surface of the earth—that is what makes them hazardous to those of us living here. Meteorological hazards take place in the lower atmosphere, and as they move through it, they can sometimes affect areas hundreds to even thousands of miles across. But these processes are self-limiting, and the rain and wind of storms serve to clean contaminants out of the air. It is the stratosphere, the part of the atmosphere higher than eight to twelve miles above the earth's surface, that protects our planet from outside radiation and unites the climate systems of the world together.

Many volcanoes stay local. The volcanoes of the mid-ocean

ridges, erupting underwater, have no impact on the atmosphere whatsoever. The nonexplosive eruptions, such as the Kilauea eruption that has been continuing in Hawaii for over thirty years, release gases that stay near the earth. Even explosive eruptions can have a limited impact.

The two most common gases by volume in most eruptions are the steam from water and carbon dioxide, already common components of the earth's atmosphere. And even explosive eruptions are rarely powerful enough to put significant amounts of material into the stratosphere. Here the Icelandic volcanoes have an advantage. The lower atmosphere is thicker at the equator than it is at the poles. Mount Pinatubo had to eject material to a height of twelve miles to get it into the stratosphere, whereas in Iceland, the material had to reach a height of only eight miles. The Icelandic volcanoes are likely to continue to have an impact on the world.

Climatic disruption from volcanoes is temporary, however. It represents a onetime injection of gases into the atmosphere, usually lasting weeks or months. Also, the most significant gases are heavier than air, and they undergo chemical reactions with other elements in the air that allow them to escape the atmosphere through precipitation. Natural circulation removes the gases in a few years, and with it their effect on climate.

The gases that humans are adding to the atmosphere are likewise affecting the climate of the planet. Unlike the sulfates in the stratosphere that block sunlight from coming in, and that consequently cool the planet, carbon dioxide and methane in the lower atmosphere block infrared radiation (or heat) from radiating back out, heating the planet. Because these gases are lighter, they do not simply precipitate out of the air like volcanic gases. Moreover, our current addition of carbon dioxide to the atmosphere through the burning of fossil fuels is ongoing, not a onetime event. The global damage inflicted by Laki offers a sobering vision of the impact not just of a single volcano, but of what can happen when our shared atmosphere is contaminated.

CHAPTER FOUR

WHAT WE FORGET

California, United States, 1861–62

I don't think the city will ever rise from
the shock, I don't see how it can.

—William Brewer, March 1862

As an earth scientist, I have learned to be comfortable with geologic time. I can talk about the last ten thousand years as "recent" without any sense of irony. I look at mountains that others see as the foundations of the earth and I see motion—faults pushing them up faster than erosion can wear them down. From this vantage point, I marvel at cities that lie within floodplains, on the flanks of volcanoes, straddling active faults. My astonishment is not at the cities being there. These features offer their advantages, as we've seen. What puzzles me is the inability of the cities' inhabitants to recognize their risk and to do something about it. To a geologist, "sometime in the next millennium" sounds not like an evasion but like a threat.

But for most of us, the future remains an abstract concept. We show a remarkable ability to forget past disasters, or to minimize their imagined impact. Try asking a Californian what the worst natural disaster was in its nearly 170-year history. A recent transplant will respond with a recent earthquake—in San Francisco (1989, Loma Prieta) or Los Angeles (1994, Northridge), perhaps. Those with family roots in California might say the 1906 San Francisco earthquake, which released fifty times the energy of the Northridge quake.

But in fact, the most devastating event in California history was a flood. In the winter of 1861–62, rain across the West created the worst flooding in the history of California, Oregon, and Nevada, killing thousands—more than 1 percent of the population at the time—and bankrupting the state. A three-hundred-mile stretch of California's Central Valley, the center of its agriculture, was covered thirty feet deep in water. And yet most Californians have never heard of it.

I grew up a fourth-generation resident of Southern California, and *I* had never heard of the Great Flood either. Not until I started a program for USGS to model the big disasters in California's future. In addition to studying earthquakes, my team wanted to create models for floods—ones that occurred with the same sort of frequency as a San Andreas earthquake, once every century or two. There is no point modeling the impact of an event that happens every thirty or fifty years. You have hard data, not to mention the records of people who lived through it; you don't need a scientist to tell you what to expect. I asked the hydrologists I was working with what the biggest storm was that they knew about and almost could not believe their answer.

*

California was claimed by Spain in the sixteenth century and named after a fictional island in a popular contemporary Spanish novel, a far-off place of abundance led by an Amazon queen, Califia. The reality of early California, before irrigation systems, was not nearly so bountiful. With three to four months of rain per year and prolonged, dry summers, it couldn't support most crops. It had drawn just a few thousand European colonists by 1821, when a revolution transferred California from Spain to the Mexican Empire. Contributing little to the tax base, California was largely ignored by the

Mexican government. Most of the *Californios* (Spanish-speaking descendants of the original colonists) could be found in the southern part of the state, and the arrival of a few English-speaking settlers in the northern part went unprotested. When California was lost to the United States in the Mexican-American War in 1848, there were fewer than eight thousand white Hispanic residents and about fifty thousand Native Americans.

This all changed when gold was discovered near Sacramento in 1848. Word spread quickly, and in 1849 settlers began pouring into California. The political leadership scrambled to handle this wave of migrants (the forty-niners), and California was quickly granted statehood in 1850. The first U.S. census that year showed just under ninety thousand residents; the population had ballooned to over four hundred thousand by 1860. In their quest to find gold and get rich quick, most failed. The more reliable path to success lay in supplying the human needs of the miners.

The primary industry, prior to the gold rush, had been ranching, selling leather and tallow to shippers to take to the East Coast. But with the influx of new residents, farms, shops, and light manufacturing rapidly developed in the northern part of the state, along with saloons, gambling houses, and brothels. The gold was shipped out through the harbor in San Francisco Bay, and San Francisco quickly became the state's largest city. The capital of the new state moved through a few locations before settling in Sacramento, center of much of the gold rush activity. The area between San Francisco and the Sierra Nevada foothills above Sacramento held four-fifths of the California population. Southern California was still dominated by the Californios and their ranches, the lack of reliable summer water restricting agriculture to just the regions near the larger rivers.

By 1861, the headiest days of the gold rush were past and the state was fleshing out its infrastructure. The legislature had just established the California Geological Survey to catalog the state's resources, enlisting Josiah Whitney to be the state geologist. The

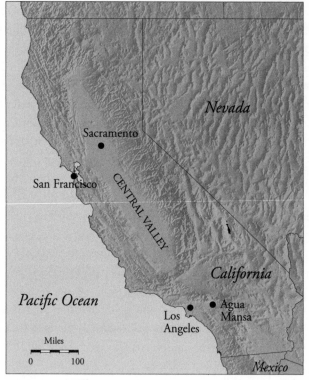

Map of California

men who had profited from the gold rush were running the state now, and they wanted to make sure they could keep the good times rolling, so they looked to Whitney, who had been part of several geologic surveys, to help them find more treasure.

Whitney had a somewhat different view of his new role. With a spectacular blindness to the political realities, he set out to discover the *scientific* treasures waiting to be understood. Whitney's first two publications, after three years of work, were on paleontology, flora, and fauna. The legislature cut his budget, and he responded by calling them corrupt, reckless, stupid, malignant jackasses. No one else was surprised when the legislature cut off funding for the survey. Most of the work Whitney envisioned never came to fruition, though some excellent reports were eventually published.

He left behind one other hidden gem, though. On Whitney's team was a young botanist named William Brewer, who kept a detailed journal of his time surveying the state. That journal has allowed us to know much more about the extraordinary winter of 1861–62.

*

Coastal California has an archetypal Mediterranean climate, where large subtropical areas of high pressure in the atmosphere drift northward in the summer, blocking most precipitation. In the winter, the cells shift southward, letting in westerlies and carrying the storms that bring rain to the lowlands and snow to the Sierra Nevada. Melting snow brings water later in the year, allowing for agriculture to develop in places that might otherwise be too dry. But some years the pressure cells linger, and the winter rains and snow quickly peter out. In other years, a wind pattern develops that brings in storm after storm, flowing across the state. Like its economy, California's climate is boom-and-bust. The 1850s had seen a pervasive drought, limiting agriculture even as so many new residents were pouring in.

When the storms do hit California, they hit hard. Everyone knows that the southern states along the Gulf of Mexico and Atlantic coast are subject to hurricanes, and most people know the names of the big ones—Katrina, Harvey, Sandy. The winter storms of California produce as much rain as a big hurricane, but without a name they are much harder to bring to mind. One measure of extreme rainfall is this: Does a storm exceed sixteen inches of rain in a three-day period? Few recording stations in the United States have ever seen this amount of rain, and most that have are in the hurricane states. Rarely has a location recorded this type of rainfall more than twice. Except in California. One location in the Sierra

Nevada mountains has recorded such an extreme rain event seven times.

Recent research has provided new understanding of these storms. In the 1990s, satellites were launched that could directly measure the water content of the air. Amazing features were revealed to us. Dubbed atmospheric rivers, narrow plumes of atmospheric moisture carry water from the tropics up into the midlatitudes. They are typically several thousand miles long and only a few hundred miles wide. When an atmospheric river comes over the coast of California, very heavy rainfall results. These storms usually persist for only a day or two. But occasionally atmospheric conditions will hold the rain door open, and flooding ensues.

Scientists have tried to reconstruct what happened in 1861–62 from the limited weather records available. It looks like the conditions for atmospheric rivers were established, and then they migrated from north to south. The rain began in Oregon in early December 1861, blasted Northern California in late December through January 1862, and put Southern California in the bull's-eye at the end of January. The heavens opened and stayed open with constant rain for forty-five days. Destruction spread across the state, killing thousands, destroying crops, fields, herds, and businesses.

Accurate assessments of the rainfall are hard to ascertain, with quantitative measurements in only a few places. The scientist William Brewer marveled over the rain:

> Since November 6, when the first shower came, to January 18, it is thirty-two and three-quarters inches and it is still raining! But this is not all. Generally twice, sometimes three times, as much falls in the mining districts on the slopes of the Sierra. This year at Sonora, in Tuolumne County, between November 11, 1861, and January 14, 1862, seventy-two inches (six feet) of water has fallen, and in numbers of places over five feet! And that in a period of two months. As much rain

as falls in Ithaca in two years has fallen in some places in this state in two months.

In Southern California, even less data is available, but one record in Los Angeles showed sixty-six inches of rain, in a region that usually sees less than thirteen inches per year.

Whatever the numbers, the result was destruction across the state. The Central Valley of California is a huge depression running much of the length of the state, between the coastal ranges to the west and the Sierra Nevada to the east. The torrents coming out of the Sierra Nevada filled the valley with water that took most of the next year to drain. The capital city, Sacramento, was built at the confluence of the American and Sacramento Rivers, in the flatlands at the north end of the Central Valley. As the second-largest city in California, it suffered the most concentrated losses.

The worst flooding arrived on January 9. The American River rose first, and its levee failed that morning. The Sacramento River had not yet crested and its levees held—trapping the water from the American River *within* the city. By January 10, the river in Sacramento was twenty-four feet above low water. Since most of the city sat at sixteen feet above low water, it was now submerged in water eight feet deep. *The New York Times* reported: "The greater part of the most fashionable houses had from three to six feet of water in the parlor. In many of the houses the line of the flood is visible on the plastering in the second story. Dozens of wooden houses, some of them two stories high, were lifted up and carried off. . . . All the firewood, most of the fences and sheds, all the poultry, cats, rats and many of the cows and horses were swept away."

The next day was inauguration day for the newly elected governor, Leland Stanford (who later founded Stanford University). Unfortunately, the new capitol sat in the middle of the floodwaters. Participants could get there only by rowboat. Concerns about safety were waved aside, and the inauguration was held as planned at the

Sacramento during the Great Flood. K Street east from Fourth Street. Stereograph from the California Heritage Collection, UC Berkeley, Bancroft Library.

capitol. The governor was sworn in and rowed back to his mansion, only to enter by steering his boat to a second-floor window. The new state government tried to function, but the support for life in the city of Sacramento was collapsing. After twelve days, the fight was abandoned and the state government moved to San Francisco.

The devastation was so pervasive that some thought it impossible to recover. William Brewer came back to Sacramento in March and recorded the scene:

> Most of the city is still under water, and has been there for three months . . . every low place is full—cellars and yards are full, houses and walls wet, everything uncomfortable. . . . Yards were ponds enclosed by dilapidated, muddy, slimy fences; household furniture, chairs, tables, sofas, the fragments of houses, were floating in the muddy waters or lodged in nooks and corners. . . . Not a road leading from the city is passable, business is at a dead standstill, everything looks forlorn and wretched. Many houses have partially toppled over; some have been carried from their foundations, several

streets (now avenues of water) are blocked up with houses
that have floated in them, dead animals lie about here and
there—a dreadful picture. I don't think the city will ever rise
from the shock, I don't see how it can.

*

Of course, Sacramento did eventually recover. It did so through
audacious vision, determination, and an extraordinary drive of
engineering. The solution, the city government decided, was to
raise the whole city of Sacramento above the 1862 flood level and
rebuild it. The residents organized a self-tax to fund the transport
of mud and sand in order to lift the ground level of a two-and-a-
half-mile stretch of the city by nine to fourteen feet. Owners cut
some buildings off their foundations and jacked them up ten feet.
Others just abandoned their first floors and filled them in. The process took fifteen years and untold costs to complete.

The potential for catastrophe had escaped the realm of the
unknowable. Residents were driven by the very real fear of losing
their city. Beyond existential threats, they recognized that the capital was movable; the possibility that the state government might
just stay in San Francisco loomed over them.

Cities have been washed away in floods throughout human history. What made this flood so extraordinary is that Sacramento was
just one of hundreds of cities and towns lost or nearly lost at the
same time. Most other cities of Northern California were severely
damaged. On January 21, 1862, before the southern part of the state
had received its heaviest rains, *The New York Times* reported, "All
Sacramento City, save a small part of one street, part of Marysville,
part of Santa Rosa, part of Auburn, part of Sonora, part of Nevada,
and part of Napa, not to speak of less important towns, were under
water." Many small towns were completely destroyed, though
it is difficult to get a complete count. The *Shasta County Courier*
reported that, just in that county, three towns lost every one of their

houses. It appears many people left the state. Twenty months later, both William Brewer and *The New York Times* reported declining population in the state.

William Brewer started referring to the Central Valley as "the Lake."

> The "Lake" was at that point sixty miles wide, from the mountains on one side to the hills on the other. . . . Nearly every house and farm over this immense region is gone. There was such a body of water—250 to 300 miles long and 20 to 60 miles wide, the water ice cold and muddy—that the winds made high waves which beat the farm homes in pieces. America has never before seen such desolation by flood as this has been, and seldom has the Old World seen the like.

Landslides tore through many of the mountainous regions, which were still populated with miners. In the face of these landslides, combined with flooding, many lives were lost, although no comprehensive count has ever been made. As with most disasters, the poor suffered the most. They live in more precarious housing and have fewer resources with which to respond. Chinese immigrants appear to have suffered the greatest losses, based on reports from Chinese support organizations in San Francisco. The death toll among Chinese communities may have exceeded a thousand.

San Francisco suffered less than Sacramento by comparison. Built at the end of a peninsula, it had the San Francisco Bay to the east and the ocean to the west, to which the rains could escape. But even there, the evidence of the onslaught was all around them. Much of Northern California is drained by rivers flowing into San Francisco Bay. The narrow outlet of the bay at the Golden Gate had such a strong outflow that ships couldn't enter. Freshwater fish were caught outside the mouth of the once-saltwater bay. That the largest city stayed intact was probably critical for the state's survival.

Los Angeles and Orange Counties, in the southern part of the state, are now home to fourteen million people, but they had fewer than fifteen thousand residents at the time, most of whom experienced flooding. Los Angeles was the largest town of the region, followed by Agua Mansa (meaning "gentle water"). With its fertile land on the banks of the Santa Ana River, which also offered irrigation through the dry summer, the location must have seemed ideal to settlers.

In average Southern California winters, the Santa Ana drains the sparse rainfall on the San Bernardino Mountains into a manageable flow. But these are some of the steepest mountains in the world, which accelerates the impact of what is known as *orographic lift*. When storm clouds rise to get over the mountains, they cool quickly, squeezing out rain. Weather stations in the San Bernardino Mountains usually record twice as much rain as the stations in the flats where Agua Mansa was located.

On the night of January 22, 1862, after four weeks of rain and more than twenty-four hours of deluge, the town of Agua Mansa was lost to a raging torrent. The church of the community was built on a rise above the town, and its priest, Father Borgotta, heard a loud noise as the floodwaters grew. Seeing the danger, he began ringing the church bell and didn't stop. The residents of the town came to the church to find out why the bell was ringing, and there they stayed, in the only place of safety. Father Borgotta kept ringing the bell even as the floodwaters continued to rise. The last few people waded and swam to get to the church.

Because of Father Borgotta's quick thinking, no one died in Agua Mansa, but the town was destroyed. Adobe houses dissolved in the rushing water, and the cropland was covered in rocky debris that came out of the mountains. Not one building except the church and its vicarage was left standing. The townspeople's herds had been carried off and drowned. The fields could not be tilled in the spring because the rocks and boulders brought down in the flood blocked their plows. The only remnant left of Aqua Mansa today is the stairway to the church that saved the lives of its parishioners.

The water remained throughout the state for months, transforming the land. It extended for four miles around the Santa Ana River in Anaheim, creating an inland sea four feet deep that lasted for a month. When the flood finally receded, the mouth of the river had moved six miles. In Los Angeles, the water was described as extending from mountain to mountain, with no dry land in the fifty miles between the Palos Verdes Peninsula and the San Gabriel Mountains, an area now home to almost ten million people.

In the "lake" that had been the Central Valley, the water persisted throughout the year. The tops of the new telegraph poles connecting San Francisco to New York were completely submerged; the tops ceased to function for months. Roads were impassable, so the mail could not get through. For a month, all communication with the outside world was lost. Each community knew what had happened to it, but it took many months for even Californians to realize what had happened to the rest of their state. The news that Southern California had suffered any damage whatsoever did not reach the capitol until late February.

Even now, it is hard to grasp the full extent of the flood's destruction, because damage to the smaller communities did not get recorded when there was more exciting damage to discuss in the larger cities. We know from property tax records that one-third of the taxable land was destroyed (and thus didn't contribute taxes in 1862). The state went bankrupt. The legislature wasn't paid for eighteen months. (Given that, we might be a bit more sympathetic to their decision to cut off funding for Professor Whitney's survey.)

The Great Flood fundamentally changed the economy of California. Whole industries were crippled. Fresh water, laden with heavy sediment, flushed over the oyster beds of San Francisco Bay, halting production. Mining equipment was washed out of the mountains, and many miners were lost; the loss of manpower and equipment marked the beginning of the end of the gold rush. The ranching industry that had defined the culture of Southern California shrank to become just a minor player. Herds were devastated by the floods, with two hundred thousand head of cattle, one hundred thousand sheep, and five hundred thousand lambs drowned. The ranchers could not afford to restock their herds, and in an ironic twist, the next two years saw a severe drought that compounded the losses. California switched from a ranching to a farming economy.

*

Just trying to describe the extent of the damage is overwhelming. This wasn't the kind of flood we think we understand—it was ruination extending across thousands of miles of inhabited land. And yet, 150 years later, most Californians are unaware that it ever happened. The floods are noted in Sacramento, but they're thought of as a local event. The disaster and recovery are held up as a testament to the resilience and ingenuity of their community. The city offers museum tours of buried first floors, called the Sacramento Underground. But otherwise, across the state, residents dwell on droughts and contemplate earthquakes, but pay little mind at all to flooding.

How could California simply forget a catastrophe of this magnitude?

Two factors, psychological and physical, have played a role in our communal amnesia. Evolutionary psychology studies the ways that our thinking and feeling have been transformed by evolutionary pressures. We evolved into human beings in a world of predators and famine, where responding quickly to short-term crises was essential for our survival. Risk was all around us, and the most successful breeders were those who learned to recognize the most imminent ones. To most of us, flooding doesn't feel imminent. Without a personal association—firsthand experience, or else the recollections of a parent or grandparent, say—our connection to disaster can become so tenuous as to lose its emotional grip altogether. And when it comes to our evaluation of risk, emotion is often more powerful than reason.

There's another, related psychological tendency at play. Flooding is always seen as more benign than other hazards, in spite of the huge death and economic toll it takes, because of the familiarity of its sources. In that prehistoric world in which we became human, the predator that could be seen was often less dangerous than the hidden one, lying in the grass. The predator you could see, you could protect against. Not so the snake hiding in the grass. We continue, accordingly, to fear those risks that lurk out of sight. We fear the threat posed by nuclear energy, even though the only American nuclear accident, Three Mile Island, killed no one, but pay little mind to the act of driving, even as more than thirty thousand Americans die in car crashes every year. We fret about cancer from cell phones as we drag on cigarettes.

Rain is so familiar as to feel benign. The floodwaters may rise, but you can see the water approaching. Its impact feels manageable. And most of the time, we manage it quite well. The other hazards—earthquakes, volcanoes, landslides—emerge from out of nowhere. They are erratic, invisible, a sudden disruption to the earth. Not so rain.

Beyond the psychological factors are the physical ones. For all natural disasters, smaller events occur much more often than larger ones, with the largest events rarest still. Look at the earthquakes recorded throughout the world this year, or across all of California's history, or even just the aftershocks of one mainshock, and you'll find the same distribution of sizes. For every magnitude 7, there will be ten magnitude 6 events, one hundred magnitude 5 events, one thousand magnitude 4 events, ten thousand magnitude 3 events, and one hundred thousand magnitude 2 events.

For floods, a similar pattern applies (though statistics are unique to each drainage system). In any river, we can measure the flow rate—how much water passes a given point in one second. Most of the time, it is low. When a typical storm hits, it goes up slightly. Every few years, a really big storm strikes and it goes up further. If it rains on a lot of accumulated snow, even greater flow results. When all the factors line up, they trigger a catastrophic flood. Just like earthquakes, the small ones are common, the big ones rare.

Hydrologists measure flow rate every day. They plot out daily rates for many years. They can say that a stream has a high probability of exceeding the low value in any given year, a moderate probability of exceeding the high value, and a very low probability of exceeding the very high value. The flow rate that has only a 1 percent chance of being reached at any time in one year is called the hundred-year flood. You can push out the distribution curve, assuming that this relationship of large to small will continue, and estimate the thousand-year flood—the flood that has only a 1-in-1,000 chance of happening this year. That's a low probability, of course; but with thousands of rivers, each receiving different storms, we see a "thousand-year" flood somewhere in the world most years.

The 1861–62 flood was the worst of many very devastating floods in California in the nineteenth century. By the early twentieth century, the great river delta of the Central Valley had been contained by levees. Dams were built in the Sierra foothills to save water for irrigation and to reduce further flooding. After one rainstorm in 1938 dropped thirty-two inches of rain in five days in the mountains around Los Angeles, putting one-third of the Los Angeles basin underwater, the demand for flood control in the southern part of the state was too great to ignore. The streams in Los Angeles were lined with concrete to hasten water's passage to the ocean. Through man's ingenuity and engineering, the floods were stopped.

Or so it seemed. What the engineers had done was manage smaller floods. The dams, levees, and concrete channels could accommodate hundred-year floods, or in a few cases, the two-hundred-year flood. But however big a flood control system you build, a larger flood is always possible—in fact, it is almost certain if you wait long enough. At some point in the future, another winter like 1861–62 *will* come along, overflowing dams and breaking through levees, flooding potentially millions of homes. It's not a question of *if* but *when*.

*

That's where my team and I stepped in. In 2008, just as we were releasing the last phase of the San Andreas earthquake scenario, we undertook the modeling of a massive flood for California. Our aim was for something like 1862, although modeling constraints made it a bit smaller. The program was called ARkStorm. The "AR" stands for atmospheric rivers, the meteorological phenomena behind the big storms, and the "k" stood for "1,000" to suggest that we were looking at rare, big storms. (In fact, the "k" was somewhat arbitrary, but it allowed us to use cool graphics of arks.) We estimated from geologic records that storms the size of 1861–62 happen once every century or two—much the same rate as the big San Andreas earthquakes.

The flooding we saw was big enough to overwhelm existing flood control systems and thus return California to flooding levels much like those of the nineteenth century, before any such controls were in place. We were astonished to see how much more damaging ARkStorm was compared to the ShakeOut earthquake. Using the same methodology wherever possible, we found that 24 percent of the buildings in California would be damaged and that the losses to flood were four times greater than for the earthquake, with almost $1 *trillion* in damages.

But even more surprising than the findings themselves was the reaction to our study when it was released in 2010. Many officials simply refused to accept it. Where the ShakeOut scenario had been embraced by emergency managers, many flood managers dismissed out of hand the possibility of that much damage. They knew what a flood was—they had managed many floods in the past. They wanted to believe that their engineering solutions could not be exceeded, so they ignored our findings. (The scenario found more traction on local levels, among smaller regional organizations. It was easier for them to imagine losing their communities.)

Which brings us back to the psychological dimension. As disaster scientists, we knew that floods cause less emotional distress than earthquakes, so we shouldn't have been too surprised. But I thought that when presented with the evidence, cities would say, "We need to change our priorities." Instead, the data was largely rejected because it didn't conform to the emotional response of the emergency managers—people who, like all of us, are more afraid of the unseen. Psychologists might say they were acting on confirmation bias, critical of data that didn't match their point of view.

This inability to accept the possibility of extreme flooding events increases the risk for people across the United States, and indeed the world. The standard method for estimating how bad a flood could be requires using recorded history. The length of that history—what the hydrologists call the "period of record"—rarely exceeds one hundred years in the United States, because the first stream gauge was invented only in the late nineteenth century. That means that old floods, such as those in 1861–62, aren't integrated into predictions, even when we know about them. Across the country, people build houses and businesses in floodplains based on insufficient data, creating greater risk with each new construction project.

The situation is potentially even more perilous today. The additional heat in our atmosphere, which has, over the last century, warmed average global temperature by 1.5°F, translates into extra energy, contributing to more extreme storms. The assumption underlying any "thousand-year flood" is that the future will be similar to the past—what scientists call "stationarity." With several thousand-year storms happening in just a decade in communities such as Charleston, South Carolina, and Houston, Texas, the phrase "Stationarity is dead!" has come to be pronounced at more than a few hydrology conferences and workshops.

When one-third of the taxable land of California was destroyed in 1862, California had four hundred thousand residents. It now has almost a hundred times more people, all still at risk from a future flood. And almost none of them know it.

CHAPTER FIVE

FINDING FAULTS

Tokyo-Yokohama, Japan, 1923

If this were not Hell, where would Hell be?

—Anonymous survivor of the 1923 Kanto earthquake

Earthquakes are as central to Japan, its history, and its culture as Mount Fuji or the emperor. With a frequency of earthquakes three times that of California's, Japan has been devastated by earthquakes, and their resulting tsunamis, repeatedly through its history. Japanese mythology holds that Namazu, a giant black catfish buried underground, is responsible for earthquakes; woodcuts depicting the earthshaker are widely made and sold. Typhoons and lightning have patron deities, but earthquakes alone have been attributed to Namazu's malevolence, tamed only by a Shinto god.

One of the deadliest earthquakes ever to ravage Japan was the 1923 Kanto earthquake of magnitude 7.9 that destroyed most of Tokyo and Yokohama and killed over 140,000 people. It occurred as Japan was transitioning from a traditional, isolated culture to a player on the world stage, and the nation's response to the earthquake reflected that dichotomy.

The Japanese Empire had maintained sovereignty over its islands for more than a thousand years, secure in the knowledge of their special place on earth. The emperor was considered a

Ureshi taian 'nichi ni yurinaosu. "How happy. The shaking rectification of the great day of peace." From the Nichibunken Archive.

semidivine figure, descended from the sun goddess and a mediator for the Japanese people with the celestial world. The shogun, officially acting in the emperor's name, was the actual ruler; the emperor had, over the centuries, become a primarily symbolic figure.

Japan's cultural and philosophical understanding of the world was influenced by Confucian scholarship from China. The oldest Chinese book, the *I Ching,* written more than three thousand years ago, is a foundational text for Chinese and Japanese philosophy alike. Confucius studied the *I Ching* and wrote extensive commentaries on it. In the Warring States Period (around 250 BC), the philosopher Zou Yan developed the School of Naturalists, also called the School of Yin-Yang, from ideas in the *I Ching.* It explained the universe as dependent on the interplay of essential forces: the yang (male, light, air, hot) and the yin (female, dark, earth, cold), acting through five elements (water, fire, wood, metal, and earth).

The ideas of the yin-yang school were integrated with Confucianism by Dong Zhongshu, an adviser to the Han Dynasty emperor, in the second century BC, creating a philosophy that would guide the Chinese imperial government for two thousand years and strongly influence Japan. His book, *Luxuriant Dew of the Spring and Autumn Annals,* described a world in which the celestial, human, and natural spheres are interconnected, with each having to maintain a balance between the opposing forces. An imbalance in the human sphere would be echoed in the natural—resulting in natural disaster.

Dong Zhongshu's writings emphasize the role of the emperor in connecting the spheres, and he developed guidelines for the emperor's behavior to stave off disasters. If the emperor was too autocratic and would not allow his ministers their proper role in the functioning of government, typhoons would strike the empire, in an excess of yang. If the emperor became too weak, if the ministers usurped the roles that properly belonged with the emperor, or if women entered government, the too-powerful yin would cause the earth to rise up and overpower the sky, leading to earthquakes. These views were so completely absorbed into Japanese culture that as early as AD 675, the Japanese government had a bureau to provide advice on yin-yang balance.

With the first arrival of Westerners in the seventeenth century, the Japanese perceived a threat to their way of life, and so in 1635, to restore harmony, the shogun prohibited any foreigners from living in Japan, on pain of death. That edict, and Japanese isolation, persisted for the next two centuries. The nation's traditional views remained entrenched, even as the Enlightenment took hold in the West. Up through the nineteenth century, it was accepted that benevolent governments would demonstrate their cosmic rightness by a lack of natural disasters, But, as the historian and professor of Asian studies at Penn State Gregory Smits noted, when "divergence between cosmic moral principles and the state of government increased, strange atmospheric phenomena, crop failures, epidemics, earthquakes, and other natural disasters became the concrete manifestations of cosmic displeasure."

Thus the Japanese, like every human society before and since, resisted the notion that earthquakes, famines, and volcanoes could strike at random times. Patterns were sought, meanings inferred. In the Judeo-Christian tradition, where individuals have a personal relationship with their God, disasters were seen as the result of individual choices to sin. In the Japanese world, social harmony and community were prized over the individual. Consequently, those same disasters were attributed to societal failures.

In the mid-nineteenth century, just as California was coping with its massive floods, Japan was struggling to fend off a wave of dangerous ideas. The isolation that had protected Japanese society for centuries came apart when, in 1853, Commodore Perry of the U.S. Navy forced his way into Japan. He sailed gunboats into Tokyo harbor, demanding that Japan open itself to trade with the West. The ruling lords were shamed by their powerlessness, their inability to oppose Perry without a navy of their own. The Japanese had no choice but to confront a radically different view of their place in the universe.

Within a decade, the ruling structure that comprised a shogun and hundreds of feudal lords was toppled, in what is known as the Meiji Restoration. The emperor was restored to governing power, no longer just a ceremonial leader. With a group of close advisers, the Meiji emperor took over foreign policy for Japan, rejecting the isolationism of the shogun before him. They saw in the humiliation by Commodore Perry the need to learn from the West. The Americans' iron warships, cannons, and guns had rendered the great fighting skills and personal discipline of the samurai irrelevant. The emperor and his advisers vowed to ensure that Japan could build their own guns and warships, that they would never again be so humiliated.

To create an industrial society that could rival the West's, the Meiji emperor understood that Japan needed to reach outside itself. He sought out young European scientists and engineers, luring them to Japan with offers of funding and students.

One of these emigrés was an English geologist named John Milne. Born in 1850 in Liverpool, Milne had studied a range of subjects, from mathematics, surveying, and engineering to geology and theology, financing some of his education by playing piano in pubs. He went to King's College and then, having decided to pursue a career as a mining engineer, to the Royal School of Mines in London. He clearly enjoyed travel, for in just a few years, he participated in exploratory expeditions to Europe, Iceland, Canada, and the Sinai. At the age of twenty-five, he accepted an offer to go to Japan, where he would have an appointment as a professor of mining and geology at the Imperial College of Engineering in Tokyo. Prone to seasickness, he traveled overland through Scandinavia, Russia, Central Asia, and China, along a route that would later be used for the Trans-Siberian Railway. He arrived in Tokyo on March 8, 1876. That night, he experienced his first earthquake.

As evidenced by his eclectic fields of study, Milne's interests were varied. He documented his observations of geology and botany during his crossing of Eurasia in his 1879 book, *Across Europe and Asia*. He studied the aboriginal people of Japan, the Ainu, traveling to the northern island of Hokkaido. While there, he met and married his wife, Toné Horikawa, the daughter of the abbot of a Buddhist monastery. (Many years later, he married her again in a church in Tokyo in a ceremony that would be recognized by the British government.) Toné worked alongside Milne as a geologist.

It was in his study of earthquakes that Milne's legacy was enshrined. Since the Enlightenment, Europeans had devoted themselves to a wide range of scientific endeavors. But with few earthquakes to observe, there weren't many researchers applying scientific rigor to the subject. It was in Japan that Milne and his fellow engineers at the Imperial College began trying to measure the earthquakes they were feeling, and in so doing created the field of modern seismology.

*

The Japanese islands are formed by the ongoing collision of four tectonic plates. The Eurasian plate (to the west) is colliding with the Pacific Ocean plate (to the northeast) and the Philippine Sea plate (to the southeast), with a sliver of the North American plate caught in between. The collision between these four plates has pushed up the two plates to the west, the Eurasian and North American plates, creating the islands of Japan themselves—not to mention the magma that forms volcanoes (such as Mount Fuji).

Subduction zones such as these have the highest rate of earthquakes of all the types of plate boundaries. They are under higher stress because the plates are pushing *into* each other—instead of pulling apart or grinding alongside each other—which allows for greater friction, with a larger percentage of the energy released in earthquakes. The complications resulting from the collision of four plates, not just two, mean that Japan's earthquakes tend to be distributed throughout the islands, not just at the plate boundary. Japan has more people at risk from earthquake shaking than anywhere else in the world.

It was here that John Milne set about his study. He was curious about the earthquakes he felt and began looking for ways to record them. After a major earthquake damaged the port city of Yokohama in February 1880, Milne, two other Englishmen, and their Japanese colleagues formed the world's first seismological organization, the Seismological Society of Japan. One of their initial steps was to develop a precision seismograph, based on a horizontal pendulum. The core principle is to suspend a large mass so that when

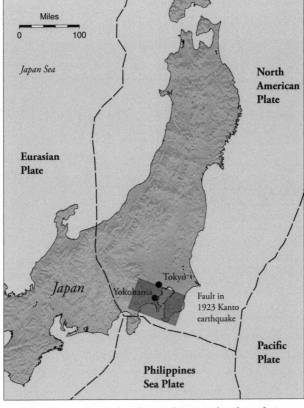

Map of central Honshu, Japan, showing plate boundaries
and the location of the fault for the 1923 Kanto earthquake

the ground starts to move, the mass will stay where it was. A pen
connected to the mass, writing on paper affixed to the ground, will
show the differential movement between the ground and the mass.
(Modern seismometers still suspend a mass, but they use an electrical feedback system to create a digital record of the motion.) With it,
the first systematic catalog of earthquake locations and sizes could
be constructed. That year, Milne's student Seikei Sekiya became the
first person anywhere to be named a professor of seismology. He

was also named head of the world's first seismology department at
the University of Tokyo.

John Milne (center) and his wife, Toné, looking at one of his horizontal pendulum seismographs with Russian seismologist Prince Boris
Galitzin. From the collection of the Carisbrooke Castle Museum, Isle
of Wight.

With all these firsts—new instruments, new data, new researchers—the field of seismology quickly flourished. But the
seismologists found, as they have continued to find, that the most
significant advances spring from the most interesting earthquakes—
unlike other scientists, seismologists are in no position to create
their own experiments.

In 1891, the magnitude 8 Mino-Owari earthquake offered fod-
der for early seismologists. The fault it occurred on was not part of
the boundary between the three tectonic plates that lay offshore
but was rather a secondary, inland fault. This fault came up to the
surface of the earth, offsetting streams and roads. Studying that
feature, Milne was the first to hypothesize a relationship between
earthquakes and faults. (He had the causality reversed, conjectur-
ing that the earthquake had broken the fault. It would be decades
before scientists realized that it was the movement along the fault
itself that caused earthquakes.)

Another student of Milne's, Fusakichi Omori, used the Mino-Owari earthquake to recognize and quantify the largest nonrandom component of earthquake behavior: aftershocks. When one earthquake happens, other earthquakes become more likely. The movement on a fault stresses everything around it, creating new irregularities, concentrations of stress. Subsequent earthquakes will result to relieve those new stresses. Because those earthquakes are expected and are a direct result of the preceding quake, we call them aftershocks.

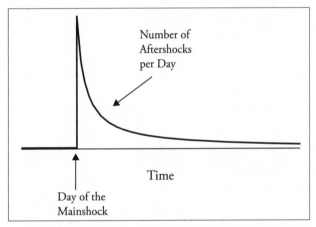

How do we know it's an aftershock? The word *aftershock* is used to describe earthquakes that occur after the main shock at a rate higher than was seen before the main shock.

Omori studied the aftershocks of the Mino-Owari earthquake and showed that their number drops off with time, and that the pattern can be described with a simple equation. If there are 1,000 aftershocks on the first day, there will be 1,000÷2, or 500, on the second day, and 1,000÷3, or 333, on the third day. The tenth day will see 1,000÷10, or 100, aftershocks, and the hundredth day will see 1,000÷100, or 10, aftershocks. This means there is a very quick drop-off, but then a very long tail. The ninety-ninth day will have 1,000÷99, or 10.101, aftershocks, so you can see that day ninety-nine and day one hundred are very similar. In 1991, on the centenary of the 1891 earthquake that Omori studied, Japanese seismologists showed that the rate of earthquakes around that fault was still following the decay pattern that Omori had discovered a hundred years earlier.

Soon after the Mino-Owari earthquake, in 1895, Milne returned to Great Britain with his wife, Toné. He built similar seismographs on the Isle of Wight, installing them in multiple locations. Communicating with his Japanese colleagues, he was able to show that some of the movements being recorded in England were caused by Japanese earthquakes. With this observation began the field of global seismology.

Back in Japan, Fusakichi Omori was named chairman of the seismology department after Sekiya's death in 1896. He expanded the staff and students. Instrumental seismology was up and running, with the Japanese leading the way.

When the great earthquake on the northern part of the San Andreas Fault destroyed San Francisco in 1906, the Japanese emperor sent aid. Professor Omori himself went to California to provide his scientific expertise and learn from the event. Because of anti-Asian sentiments in California, he was mistreated when he arrived, even assaulted on the street. It is fortunate for all of us that he persisted. His work with scientists at the University of California, Berkeley, helped bring the pursuit of seismology to the United States.

*

Milne had connected the occurrence of earthquakes with faults that could be seen at the surface. A decade later, a student of Omori's, Akitsune Imamura, tried to further quantify the spatial patterns of earthquakes. Luckily for him, the spatial distribution of earthquakes is much more predictable than the time of their occurrence. Going over Japanese history, Imamura saw a pattern of repeating earthquakes under the Tokyo-Yokohama metropolitan area. He worried about what a major earthquake would do to the newly urbanized area.

When an earthquake moves on a fault, every point on the fault's surface produces shaking. If the fault is vertical, as with the San Andreas, the concentration of shaking might be represented on the earth's surface as a line. The shaking generated on the fault radiates outward from that line, away from the fault, dying off with distance. We therefore see a linear pattern of shaking. If the fault is more horizontal than vertical—the case with the main fault in subduction zones—then the area of strong shaking will be wider. It might be represented on a map not as a line but as an area, and it can encompass an entire city.

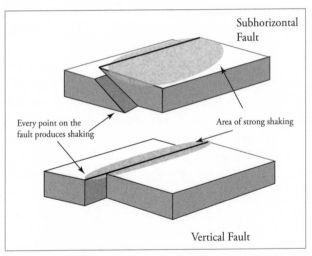

Comparing earthquake shaking around a vertical and a subhorizontal fault

Imamura saw that previous earthquakes had shaken all of what was now greater Tokyo and Yokohama. The region had since grown rapidly to a combined population of four million people, supporting the nation's new industrialization program. More than half of the inhabitants had arrived in the preceding two decades, with many living in crowded, poorly constructed buildings. Much of the land was soft sediment, and scientists had already recognized that earthquake shaking was worse in such soils. Imamura realized that a very large earthquake in such conditions would be devastating. He particularly worried about the fires that could be triggered and potentially grow too large to fight. Both of the very big earthquakes that had hit large, dense cities up to that point—the 1755 magnitude 8.7 Lisbon earthquake and the 1906 magnitude 7.8 San Francisco earthquake—had produced firestorms large enough to wipe the cities out.

Imamura knew the history of these other earthquakes and saw the potential for a similar conflagration in the packed-together wooden houses of the new Tokyo. He prepared an analysis estimating the impacts of such an earthquake and the resulting fires, publishing it in a relatively obscure journal in 1906, just after the fires that had destroyed San Francisco. He estimated that between one and two hundred thousand people would die. A popular Tokyo newspaper took up the story, trumpeting the most sensational parts and proclaiming, without his permission, "Imamura Says Tokyo Will Be Destroyed." This generated public fear, and Omori, the director of the seismology department at the University of Tokyo, was furious. He publicly castigated Imamura and wrote his own article explaining why Tokyo would not suffer a major earthquake for centuries. The public humiliation outraged Imamura. He and Omori rarely spoke again.

Education in science during the Meiji Restoration had helped shift the general understanding of earthquakes as physical phenomena, rather than spiritual manifestations, just as the development of natural philosophy in late-eighteenth-century and nineteenth-century Europe helped spark a similar evolution in thought. But for both cultures, folk traditions and emotional aversions were not so easily overcome.

The Meiji emperor passed away in 1912, and his son, the Taisho emperor, ascended the Chrysanthemum Throne. The Taisho emperor had a history of significant physical and neurological ailments, having contracted cerebral meningitis soon after his birth. His condition continued to deteriorate throughout his reign, and he wasn't seen in public after 1919. By 1921, his son Hirohito was named the prince regent, taking on his father's duties. In August 1923, the government entered an even more uncertain stage with the death of its prime minister.

In his classic Chinese text, *Luxuriant Dew of the Spring and Autumn Annals,* Dong Zhongshu had given specific advice to the Han emperor about the mistakes that would create an excess of yin energy and cause earthquakes. The most common was the impending death of the emperor. To the Japanese, for whom the shogun had filled the leadership role of the emperor for centuries, both emperor and prime minister could be seen as sources of yang energy. A feeble emperor and a dead prime minister would have seemed a precise enactment of the causes of earthquakes.

*

On September 1, 1923, at 11:58 a.m., an earthquake began that, in every sense, shook the foundation of Japanese society. A subhorizontal fault lying under Tokyo and Yokohama began to move. The rock above the fault surface, an area about forty miles wide and eighty miles long, jolted to the south. Unlike an earthquake on a vertical fault, every location situated above the fault area was literally *on* the earthquake. Yokohama sat atop the fault, with Tokyo just beyond it. So all the four million people of these two cities received some of the strongest shaking the earth can produce.

Many of those people had returned to their homes for the noon meal and were cooking over open-flame stoves. The stoves toppled and fires started, just as Imamura had predicted sixteen years earlier. Imamura himself was at his office at the University of Tokyo. As his building shuddered and roof tiles cascaded down, he did as all seismologists would in the same situation—he pulled out his

watch and timed the arrivals of different types of shaking. Later
analysis has shown that the earth produced energy for about forty
seconds, as the fault break that had begun at the western edge of the
fault worked its way east. The seismic waves would have bounced
around with a sort of echoing effect, causing shaking in Tokyo for
even longer. And that was before the aftershocks began. In great
earthquakes of this size, it feels like the shaking goes on forever.

Within minutes of the initial earthquake, fires had started across
Tokyo and Yokohama. Large aftershocks in the first ten minutes,
some above magnitude 7, impeded efforts to control the flames.
People tried, but as the fires raged, they were left with no choice but
to flee. This mass exodus clogged the streets. Survivors told stories
of being trapped for hours, unable to move. A house of courtesans
had prevented the women from fleeing, afraid they would run away
and not return, leading to more than one hundred women burning to their death within. Many drowned in the Sumida River after
jumping in to escape approaching flames. More than forty thousand people took refuge at a large open space at the Honjo Military
Clothing Depot.

Intense fire can create its own atmospheric conditions and
windstorms. Called fire tornadoes (or, translated from Japanese,
dragon twists), the heated air rising off a fire hits turbulent wind
conditions, forming a whirlwind of fire that spreads the destruction faster and farther. Dragon twists were spawned across Tokyo.
One bore down on the Honjo Clothing Depot. Only two thousand
of the forty thousand huddled within survived. Many were burned
alive while others suffocated in the superheated, deoxygenated air.

An anonymous description of this destruction asked, "If this
were not Hell, where would Hell be?"

The final tally suggested almost complete destruction of the
metropolitan area. The imperial family was spared because they
were not in residence at the time of the earthquake. But for everyone else, their lives were torn apart. In Yokohama, more than 80
percent of the buildings were destroyed. In Tokyo, about four hun-

dred thousand buildings, including the houses of 60 percent of the population, were lost. The death toll was at least 140,000 people.

Faced with this level of destruction, the government leader of a traditional Japan would have taken personal responsibility. Resignation, perhaps even seppuku, would be considered. The original Confucian classic by Dong Zhongshu provided directions for how the emperor should issue a self-criticism after a disaster and begin the process of correction. Almost immediately after the Ansei earthquake of 1855, which also damaged Tokyo, *namazu-e* (catfish picture prints) began to appear in hastily printed anonymous broadsheets. These were populist attacks on the government, accusing them of causing the earthquakes, and they contributed to the shogun's downfall. But with the 1923 quake, the prime minister had died of natural causes just a week prior. The emperor hadn't been seen in public for four years owing to his ill health. A weakness in the masculine side of government would have been readily apparent, and yet there was no leadership left to assign themselves the blame.

Japan had, to an extent, progressed beyond its strictly traditional views. The nation was quickly evolving into a modern industrial society, with many of its citizens having by now been educated in the sciences. Seismology remained a rapidly developing field. And so two opposing views of the earthquake were at odds—that it was caused by an imbalance of yin-yang forces and that it was a consequence of geologic factors. Clusters of the society would likely have responded differently depending on their education and background.

Admiral Yamamoto Gonnohyoe had been asked to form a government after the death of his predecessor on August 24, but that government was still in discussion when the earthquake hit. He was installed on September 2, the next day. With the devastation of the cities nearly total and confusion abundant, he and others in government could surely see that it wouldn't take much for an uprising against the government to begin.

*

In the face of loss and failure—no less a failure of this magnitude—
we often turn to blame. We have a deep-seated aversion to having
our missteps exposed, and we look for ways to avoid it. Blame provides an emotional outlet. It can also, in the wrong hands, be used
as an explicit ruse to divert attention away from oneself.

Because of Japan's self-imposed isolation from the outside
world, foreigners—*gaijin,* or "outside persons"—held a status in
early-twentieth-century Japan that could be described as less than
human. Of foreigners, the Koreans and Chinese had the most contact with Japan, with more dignity afforded the Chinese, perhaps
because of the influence of Confucianism and Daoism on Japanese
culture. The Koreans, meanwhile, had been subjected over many
centuries to raids by both Japanese pirates and more official Japanese seamen, finally were conquered and colonized by Japan in 1910.
As a colonial conquest, Korean workers were brought over to Japan
to staff the demands of modernization, but they were denied a path
to Japanese citizenship. Citizenship in Japan is carried out through
patrilineal lines, recognizing each Japanese child as a descendant of
the emperor. Koreans had no place in this scheme.

As the leaderless government struggled to respond to the
destruction of their capital and fires continued to rage across the
city, both the government and its citizens turned on minority Koreans. Within hours of the earthquake, rumors had begun to spread
that Koreans were planning an uprising. They were said to be setting
fires, poisoning wells, raping, and looting. These messages spread as
far as the island of Hokkaido, five hundred miles to the north.

Many citizens wasted no time responding. Vigilantes, called
jikeidan, armed mostly with makeshift weapons such as bamboo
spears, carpentry tools, knives, and broken glass, attacked their
Korean neighbors. On September 2, the newly installed prime minister declared martial law, moving troops into the affected region.
Survivors tell of army units pulling Koreans off a train leaving the

city and slaughtering them on the spot. At a minimum, police forces connived in the extermination; in some cases they actively participated. They moved to round up Koreans and confined them. It was justified as "protection and apprehension," though many of the people detained were then killed by vigilantes, in some cases inside police stations.

Evidence shows that the government had an active role in spreading the misinformation. The Department of the Interior sent cables to its local branches saying that Koreans were committing arson and ordering them to be rounded up. Police reports from different precincts included claims that Koreans had sparked fires with bombs, that Koreans were poisoning wells, that three thousand Koreans were looting and destroying Yokohama and coming for the capital next.

The massacre went well beyond overzealous protection. Many of the Korean victims were tortured. Corpses were found dismembered, with eyes and noses cut out, thousands of lacerations, and genitals removed. Observer accounts suggest that such tortures were often inflicted before death. In one location, as the historian Sonia Ryang, herself a Korean who grew up in Japan, wrote, "the mob lined up children in front of parents and cut their throats; they then nailed parents to the wall by wrists and ankles and tortured them to death." The attacks took on the qualities of ritual sacrifice; by torturing the outsider, Japanese society could purge itself of the flaws that had engendered the earthquake.

On the part of the government, officials may have been driven by other subconscious fears. Whether for having failed to heed Imamura's warnings or for disturbing the yin-yang balance, the blame could just as easily be laid at their own feet. Immigrant Koreans offered both a scapegoat and a diversion. That said, we cannot know if anyone in power explicitly fostered anti-Korean attacks to divert anger away from themselves. The "government" is itself individuals—people who had also lost their homes, witnessed the fires, felt the terror of uncertainty, panicked as their city was being destroyed around them. No one makes the best, most rational decisions in such a situation.

Whatever their initial motivation, it was not until late on September 3 that the police affairs bureaus notified news outlets that the previous reports of rebellion by the Koreans were unfounded. On September 4, the police sent out notices that Koreans need not be attacked to protect the city, but the damage had already been done. By September 5, six thousand of the twenty thousand Koreans living in Tokyo and Kanagawa had been tortured and murdered in what came to be called the Korean Massacre.

At its core, the massacre was a particularly violent manifestation of our rejection of randomness, of our need, when the inexplicable strikes, for us to find and assign blame. That the responsibility was assigned to a minority population may reflect as much the national ambivalence about a new and changing world as it does human nature. Science had begun to undermine prevailing theories about the causation of natural disasters, but it hadn't yet offered a satisfying alternative to the traditional yin-yang view. It was the gulf between those two worldviews that allowed the denial and anger of grief to well up and, lacking another outlet, led government and citizens alike to victimize the most vulnerable among them.

CHAPTER SIX

WHEN THE LEVEE BREAKS

Mississippi, United States, 1927

Mean old levee taught me to weep and moan.

—Kansas Joe McCoy and Memphis Minnie,
written in response to 1927 floods

The Mississippi River is the great river of the United States—so great, in fact, that its scale can hardly be overstated. It has the third-largest watershed in the world, gathering the rain and snow that fall on thirty-two American states, covering 40 percent of the territory of the United States, as well as two Canadian provinces. Its network of main stem and tributaries shuttles water to the Gulf of Mexico. Its Missouri River "tributary" is much longer than the main stem, but the more easterly branch was given the name Mississippi to more conveniently mark the boundary between English and Spanish territories.

Long before the European territorialists assigned names, the Mississippi had been the larder and highway for the original settlers. Fish and mollusks were harvested from its waters, and trade flourished along its banks. The Europeans, when they came, were much more interested in the transportation potential of the river. When French explorer Sieur de la Salle claimed the river, he saw the potential to connect France's Gulf of Mexico settlements with

Canada. But the French claim was never firm, and the Louisiana Purchase completed the transition of the river's control to America.

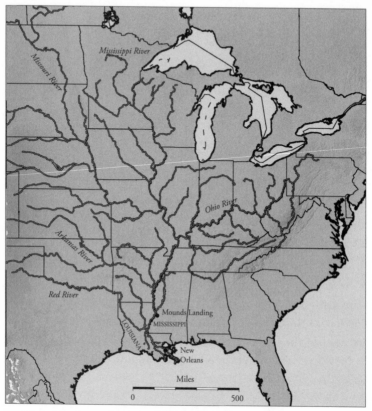

Map of the main tributaries of the Mississippi River

Thousands of miles long, this massive body of water has always been both the soul and the bane of the people who live near it. It has been the economic engine for the agricultural and manufacturing boom of the central United States. The original industries of lumber and the fur trade shipped down the river to New Orleans and out to Europe. The great, productive farms of the prairies that fed a nation and more were viable only with an efficient transportation system to take their products to our cities. Waterpower from the river ran the earliest manufacturing plants. The Mississippi became the visible symbol of the culture and lives within its region, flowing through books and songs from Mark Twain to Tennessee Williams, from Kansas Joe McCoy and Memphis Minnie to Stephen Foster to Allen Toussaint.

But riverine flooding periodically takes back the economic bounty the river provides. The history of the Mississippi is a history of flooding. The story of the explorer Hernando de Soto in 1543, written by Inca Garcilaso de la Vega in the sixteenth century, includes an account of a forty-day flood of a Native American settlement near modern-day Memphis. The nineteenth century offers a chronicle of floods, great floods, and *the* Great Floods. The songs of the river, from Johnny Cash's "Five Feet High and Rising" to Charlie Patton's "High Water Everywhere," are songs of loss and death by the waters of the Mississippi.

The success of European communities along the Mississippi has always, consequently, depended on the success of the levees around it. These embankments, both natural and artificial, help contain the river's periodic swelling, allowing life to resume as normal for the millions of residents along the Mississippi floodplains. Considering their crucial role in the area's existence, it is worth examining how natural and artificial levees form and function. To do so, we need to give up our illusion that there is an absolute boundary between water and land; between where the river flows and where it doesn't; and even between the fluid state of a river and the solid state of land.

Because we live on land, we tend to make an existential distinction between what appears at the surface and what is submerged beneath our rivers, lakes, and oceans. But a river is not fundamentally different from the land around it. There is no distinction in the types of earth that form the riverbanks, nothing unique about the crust that lies beneath it. It is simply lower lying than the areas around it. Water flows down to wherever gravity pulls it, so, inevitably, bodies of water form at those areas of lower elevation. This is stating the obvious, but it's a truth we tend to lose sight of. The Mississippi River is where it is because the land beneath it is at a lower elevation than nearly all the land surrounding it. In fact, for the last 450 miles of the Mississippi, the riverbed is actually *below* sea level. (As it approaches New Orleans, the bed of the Mississippi is as low as 170 feet below sea level.) The upper parts of the water continue downhill and with friction pull the rest of the water with them. It makes for a turbulent passage.

The amount of water contained within a river fluctuates with rainfall, and because the Mississippi drains such a large area of land, its level is dependent on rain and snowfall in many different locations. The river can be seen not as a body of water confined between two distinct lines, but almost as a living thing that swells and shrinks. Those distinct lines on a map are the river only when it is in its shrunken state, in the calm between storms. The reality of a river is whatever land is needed to accommodate its flow.

The second mind shift we need to make is to realize that a river is not just water. Moving water has the energy to carry many things within it. Naturally, the smaller and lighter the object, the easier it is to carry, and the faster the water is moving, the more it can propel. It's no accident that the Mississippi's nickname is the Big Muddy. Like all rivers, it picks up grains of sand and dirt with its motion, carrying them in suspension into the ocean. When the water slows down, it drops some of that sediment, with the bigger and heavier grains being deposited first.

Putting these two ideas together, we see why natural levees appear. When water is channeled by gravity into a river, it often carries with it a lot of mud and silt. Rain and snowmelt cause the water level to rise, such that the river is now at a higher elevation than the water ahead of it, which means it has farther to fall, and the current increases—allowing it to carry even more silt. When the water level rises high enough to overflow the riverbanks, it will continue to flow downhill, but now it is also doing so outwardly, along the surrounding land, and not just in a channel toward the ocean. No longer flowing as steeply downhill, it slows down, dropping its suspended grains of silt. The bigger grains fall first, near the river, and the finer-grained material ends up farther from the original riverbed. The result is a natural levee made up of the larger grains of sand, elevating the river's banks and rendering future floods less likely.

But as the centuries of flooding in the Mississippi demonstrate, there is always a bigger surge of water ahead, one that can overcome the natural levee (or even the artificial one). The people who have taken advantage of a levee, natural or artificial, to plant crops or build buildings in the floodplain, who have assumed the levee will always hold—or at least, hold as long as they are living there—now face a startling reality. And when flooding strikes, it doesn't strike an area or its residents equally.

How sediment in a river creates natural levees

*

New Orleans was the first attempt to create a European settlement within the floodplain of the Mississippi River. Fourteen city blocks were laid out in 1718 by Jean-Baptiste Le Moyne, Sieur de Bienville, with drainage ditches along each block to house the garrison that would control what ships entered the Mississippi. It didn't take long for residents to recognize the danger posed by the river. Within a year, flooding led Bienville to order the creation of the first artificial levees along the river—extra-compacted dirt packed three feet high. Over the next two centuries, the levees proved insufficient as floods came through the city. More dirt was added, stronger materials were incorporated, and the levees grew higher. Settlements grew north along the river, and the levees grew with them. By the mid-nineteenth century, more than a thousand miles of levees had been constructed to contain it.

Meanwhile, the field of engineering had begun to grow into a discipline, applying physical laws and calculations to the construction of our essential structures. Engineering as a concept has been part of human existence since the pyramids, but the more formal subfield began mostly as a military discipline. The first American engineers were with the Army Corps of Engineers, established in 1802 and given the charge of founding and operating the military academy at West Point. Civil engineering was later defined as those activities that were *not* military. Civil engineering schools contained within universities were established in the nineteenth century. These newly minted engineers came to the Mississippi River determined to take control.

Floods are unique among hazards in that, in coping with them, we must balance the need for containment with our other essential uses for water. Floodwater must be disposed of, but it must also be preserved for dry times (only more so in the arid West), while also retaining access to rivers for the transport of goods. (No one needs to bottle up earthquakes or magma to sell next summer.) In flooding, our most ubiquitous hazard, our need for protection can be at cross-purposes with our other economic necessities.

Since the mid-nineteenth century, engineers on the Mississippi River argued over these objectives, trying to both protect the land from flooding and keep the river open for navigation. They fought over approaches and objectives, in a dispute complicated by the rivalry between the established military engineers in the Army Corps of Engineers and the newly developing cadre of civil engineers.

The historian John Barry, in his book *Rising Tide,* chronicles what became a very personal conflict between General Andrew Humphreys, a West Point engineer and eventual chief engineer of the army, and James Buchanan Eads, a civil engineer whose entire professional life had been committed to the Mississippi—primarily to modifying it to better suit navigation. Humphreys seemed driven by vindictiveness and territorialism. In spite of his own report that made it seem impractical, he came down firmly on a "levee-only" approach, and he pulled the Army Corps with him. Their idea was that by keeping all the water *within* the river with high artificial levees, they could speed up the flow of water. The faster water would carry more sediment and could scour out the sediment that had already been deposited, removing sandbars. The floods would be stopped by the levees and shipping channels could stay open.

James Buchanan Eads had an intimate knowledge of the Mississippi few could rival. To run a salvage operation, he had developed a diving bell and had literally walked the riverbed for years. He didn't believe that the levees would provide the scour needed to clear out the sediment that blocked navigation. He pointed out that they were built quite far back from the main part of the river and confined the water only when the river was in flood. He wanted to build jetties near the mouth of the Mississippi to focus the water flow on the area of recurring sandbars. He prevailed to a degree, but because of Humphreys's opposition, these jetties were built only when Eads promised to cover the cost if they failed. Both men opposed reservoirs and dams that could reduce the water flowing into the river during floods.

By the late nineteenth century, Congress had tried to call an end to the fighting by establishing a Mississippi Commission that involved both military and civil engineers, and that was supposed to bring science to bear on the proceedings. But this political approach ended in a very political solution that turned out to be wrong in most particulars. It rejected reservoirs, which would have kept the water out of the river to begin with, and it rejected spillways and outflows, which would have allowed water to be diverted in the case of floods. It put all its eggs in a basket of levees. And for a while it worked.

In retrospect, all parties could have heeded Mark Twain's admonition when he said, "The Mississippi River will always have its own way; no engineering skill can persuade it to do otherwise." The principle that faster flow would scour sediment was based on a reasonable theoretical model. But the reality is that the Mississippi is a complicated system, and this reasoning fell apart in the details. First, because of the depth of the river, with much of its bed below sea level, the current wasn't uniform. The top of the water column was being pulled downhill by gravity, but the bottom wasn't. Friction between the water and the riverbed further distorted the currents. Second, the meanders in the river meant that the water moved faster on the outside of its curves than it did on the inside, leading to scouring on one side of the river and sediment deposit on the other. The excessive scour on some sides could actually start moving out the dirt on which the levees were built, undercutting them and, in the extreme, leading to their failure. Considering the amount of water that routinely came down the river, it's a marvel they held as long as they did.

*

The rains began in August 1926. Heavy rainfall damaged the harvest across the upper Midwest, from Indiana to Kansas, Illinois, and Nebraska. Floods spread through towns, drowning people, breaking pipelines, flooding crops. The rains continued into October, usually a relatively dry time, and Illinois and Iowa recorded flood stages higher than ever seen before. Precipitation continued into the winter. The U.S. Weather Bureau reported that on the three great rivers of the Mississippi drainage—the Ohio, Missouri, and Mississippi—every river gauge they had was recording the highest levels they had ever seen. On Christmas Day, 1926, floods ran through Chattanooga and Nashville, two Tennessee cities on different rivers. The rains continued unabated. Five different storms, each bigger than any other storm of the previous decade, struck the area around the lower Mississippi. In January, Pittsburgh and Cincinnati flooded. In February, it was Arkansas's turn as levees broke on the White and Little Red Rivers and flooded five thousand people out of their homes. The storms in March spawned tornadoes that killed forty-five people in Mississippi.

Eventually, the man-made structures on which engineers for the last century had based all their defenses began to give way. Spring represents the most dangerous time for river flooding, when melting snow adds to the rain. In the spring of 1927, so much water came into the lower Mississippi, below the confluence with the Ohio and Missouri Rivers, that the water itself began to act like a dam. The crests of floods moved more slowly, like cars compressed in a traffic jam. This only increased the pressure on levees. In the lower Mississippi area, this included not only the national "mainline" levees that had been constructed by the Army Corps of Engineers and were designed to control the Mississippi and its larger tributaries, but also the local and state levees covering smaller tributaries.

All were overseen by levee boards. Created by local and state governments, these boards often held the power of taxation, and they were responsible for local maintenance. The 1879 act that had created the Mississippi River Commission recognized the local boards and their role in keeping the system functioning. And so, as the crests approached, the levee boards stood up for the fight.

The mainline levees were massive structures. Standing two or three stories high and constructed from compacted earth, they were set back from the main river channel by as much as half a mile or more. They were built at a three-to-one slope, which means that a levee that stood thirty feet high at its highest point would be buttressed on either side by a slope that spanned ninety feet in length, with a minimum of eight feet of width at its highest point. They were so vast as to seem impregnable.

But the risk the levees faced was twofold—the pressure of the river and the malevolence of the humans in their vicinity. The confined river put immense pressure on levees, and an at-risk town's best defense could be to sabotage the levee on the other side of the river. Once one side failed, the pressure on the other was eased. So, in a real-life prisoner's dilemma, a community that was threatened and had the means, the desperation, and the shamelessness could stay safe by putting its neighbors underwater.

Thus the first and foremost function of the levee board patrols was to find and stop any saboteurs who wanted to protect their community at the expense of others. Close to a dozen people up and down the river were shot and killed by guards. Some of them may have been inadvertent casualties, but several were found to have explosives on them. Of the levees that did fail, it was hard to tell after the flood had passed whether there had been a human hand in the damage.

Sitting at the end of the Mississippi, where it meets the Gulf of Mexico, New Orleans had witnessed the failures that had taken place farther upstream, demonstrating to the city fathers the severity of the floods. In late April, the city's levees started to show signs of imminent failure. New Orleans had both the resources and the arrogance to, without stealth, dynamite the levee in St. Bernard Parish, a few miles to the east of the city. More than ten thousand residents were flooded out of their homes in St. Bernard and Plaquemine Parishes. (New Orleans later proposed a pool of $150,000 as a compensation fund for their subterfuge—less than $20 per victim. They eventually paid out several million—and even that they considered a good deal for keeping the water out of their city.)

In addition to guarding against saboteurs, the second function of the levee boards was to look for natural leaks and shore up weak spots. This meant backbreaking labor, shifting dirt at a time when there wasn't much mechanical equipment available to help.

It is here, more than in the effects of the natural disaster itself, that the greatest atrocities of the flood of 1927 began to be exposed. The rich soil of the Mississippi floodplain had been the center of the Old South's cotton plantations, and by the 1920s, with the exception of the overseers, the labor force was still completely African American, working in conditions not much different from slavery. That winter, several farmers in Louisiana had kidnapped a family of African Americans at gunpoint and taken them to Mississippi, where they were sold for $20. The victims were forced to work without pay for weeks, watched by armed guards. The white farmers were eventually indicted, but their egregiousness is telling.

When labor was needed to shore up the levees, plantation owners were asked to send their African American tenants to do the work. When that labor was insufficient, African American men were conscripted off the street, often at gunpoint, to fill out crews. As the Mississippi rose in the late winter, the levees were manned by white foremen with guns, looking for saboteurs and making sure African American laborers stayed on the job.

Publicly, the levee boards and the Army Corps of Engineers continued to insist that their structures were up to the task. But internal reports indicated that they knew otherwise. The Weather Bureau (the predecessor to the National Weather Service) said, "There was needed neither a prophetic vision nor a vivid imagination to picture a great flood in the lower Mississippi River the following spring."

As winter turned into spring and the most dangerous times for flooding arrived, more and more men were conscripted to man the levees. By the middle of March, the Mississippi National Guard had been called out to defend them. Levees had broken on three main tributaries, the White, Red, and St. Francis Rivers. By early April, over a million acres of land was already underwater.

The beginning of the end for the defenses of the lower Mississippi came with the rainstorm of Good Friday, April 15. Fifteen inches of rain fell in New Orleans in just eighteen hours, in a storm that extended over the whole lower Mississippi. Laborers were kept "topping" the levees, piling up sandbags to add to the height. On April 16, at Dorena, Missouri, a 1,200-foot stretch of the mainline levee finally crumbled, and 175,000 acres were flooded. More levees failed in the next few days.

The worst came at Mounds Landing near Greenville, Mississippi, on April 21. The levee was trembling as the water started to percolate through it. African Americans working on the levee realized it was failing and tried to leave, but they were forced back at gunpoint. When the failure finally came, many of them were swept away to their deaths. Indifferent to African American fatalities, the Red Cross officially reported just two deaths at that break.

In his *Rising Tide,* John Barry cited contemporary newspaper reports. "Thousands of workers were frantically piling sandbags . . . when the levee caved. It was impossible to recover the bodies swept onward by the current at an enormous rate of speed," wrote the *Memphis Commercial Appeal.* "Refugees coming into Jackson last night from Greenville . . . declare there is not the slightest doubt in their minds that several hundred negro plantation workers lost their lives in the great sweep of water which swept over the country," wrote the *Jackson Clarion-Ledger.*

The failure of the levee at Mounds Landing created a *crevasse* (the term for a break caused by erosion) that inundated the Mississippi Delta. The water poured through it at a rate twice that of the Niagara Falls. Within days, a million acres were submerged under ten feet of water. Many people died immediately of drowning, but most escaped to high ground. In numerous cases, the highest ground was atop the levees themselves—those that hadn't failed. Thousands crowded onto the tops of these strips of land, each no more than eight feet wide, surrounded by water, the Mississippi to the west and their flooded homes to the east.

Flooded farms covered an area fifty miles wide and a hundred miles long. More than 180,000 people lived in the region; almost 70,000 ended up in refugee camps. The headline of the *New Orleans Times-Picayune* screamed, "For God's Sake, Send Us Boats." Boats came, and the ugliest parts of the community's soul were bared.

The majority of the Mississippi Delta's residents were African American, living in conditions that hadn't much improved in the decades since abolition. After the Civil War and Reconstruction, the whites of the region had reasserted their authority through sharecropping and Jim Crow laws. Unable to vote or own land, African American tenant farmers occupied a tragic position—in debt to plantation owners, treated as less than fully human, but their underpaid labor so critical to a plantation's economic functioning that its owner did the utmost to ensure they couldn't leave.

In Greenville, Mississippi, the city in the direct path of the Mounds Landing crevasse, the head of the local Red Cross, William Alexander Percy, was aware of the horrendous conditions on the levees. He begged for whatever boats could be mustered to evacuate blacks and whites alike. His father, Senator LeRoy Percy, and other white leaders overruled him. When the boats at last arrived, only white families were allowed to board. The African Americans were left behind—without clean water, without food, without protection from the continuing rain.

*

In Washington, President Calvin Coolidge had been doing his best to keep the federal government out of the "local affairs" of the flooded midsection of the country. He had for months ignored calls for help until the crisis at Mounds Landing threatened to overwhelm residents of the region and he could ignore it no longer. Five of the affected governors had been calling for the secretary of commerce, Herbert Hoover, to be put in charge of a special federal rescue effort.

On April 22, the day after the failure at Mounds Landing, Coolidge held a cabinet meeting and agreed to this request. A quasi-governmental commission was created, which included five cabinet members and the vice chairman of the American Red Cross.

Hoover had come to public notice because of his humanitarian relief work in World War I. A geologist trained in the first class to graduate from Stanford University, Hoover had made a fortune in mining, especially in Australia and China. He was living in London as a mining engineer and financier when the Great War began and thousands of Americans were stranded in Europe with most of their traveler's checks and other financial assets unusable. In response, Hoover organized the American Committee, lending money and arranging for passage home. From there he expanded his efforts, leading the Commission for Relief in Belgium, bringing food to the civilian population caught between the armies of the great powers. When America entered the war, President Wilson asked Hoover to lead the U.S. Food Administration, which managed to successfully maintain the nation's food supply throughout the conflict. Afterward, his organization helped feed millions across Europe. By the end of the war, he was being hailed as "The Great Humanitarian."

Hoover found that making money no longer interested him—he was making more money, he said, than "probably was good for anybody." He was well known because of his war work and courted by both the Democrats and the Republicans. He chose to run for the Republican nomination for president in 1920, but his campaign fizzled. He had been so long outside the country that he didn't have a strong enough constituency to support his candidacy. He threw his support behind Warren Harding, who rewarded Hoover with a position as secretary of commerce. After Harding died, Hoover remained at Commerce under his successor, Calvin Coolidge. He stayed visible to the public through his media campaigns, but there was only so much interest he could generate in the subjects of regulating radio frequencies or conferences on street traffic. By early 1927, most articles considering possible presidential contenders for the 1928 election didn't even mention him, and those that did commented on how much the Republican establishment disliked him.

Leading the flood relief efforts in the spring of 1927 allowed Hoover's strengths of management, engineering, and humanitarianism to shine. This was fifty years before the Federal Emergency Management Agency would be created, an organization devoted to, among other things, the distribution of funds. President Coolidge absolutely refused to consider federal funding of the effort. This was consistent with a long-held view in the United States that disaster relief ought to be a local, even a personal, issue, and that it was inappropriate for the federal government to take money gathered from everyone and spend it on the needs of just a few. In

1886, President Grover Cleveland had vetoed a bill to help farmers in drought-stricken Texas, saying, "I can find no warrant for such an appropriation in the Constitution; and I do not believe that the power and duty of the General Government ought to be extended to the relief of individual suffering which is in no manner properly related to the public service or benefit. . . . [T]he lesson should be constantly enforced that, though the people support the government, the government should not support the people."

The vehicle for disaster relief at the time was the American Red Cross. In recognition of its importance, it was ceremonially chaired by the president of the United States. In 1926, President Coolidge lauded the work of the Red Cross, saying its "aid is given freely . . . and in such a way that the benefactor does not feel himself an object of charity. He does not lose his self-respect." He further emphasized, "The normal state of the American people, the standard toward which all efforts are bent for attainment, usually with success, is that of a self-supporting, self-governing, independent people." The only appropriate source of relief funds, he suggested, was philanthropic donations.

And so, the day after the Mounds Landing levee failure, President Coolidge called for donations to the Red Cross. Five million dollars was received in short order. The Flood Relief Committee that Hoover was asked to chair was to be quasi-governmental, with the Red Cross involved in its coordination.

But over twenty-six thousand square miles of land had been flooded, displacing more than six hundred thousand people and requiring a bigger relief effort than had ever before been seen in the United States. It was quickly clear that the scale of it could not be adequately addressed by normal donations. Still, the president persisted in his unwillingness to permit federal funding; he even refused to call a session of Congress to discuss it. So to raise money, Hoover turned to his skills with the media, developed while serving as secretary of commerce. He established campaigns to expose southern suffering to residents in the North. It worked; donations to the Red Cross increased to over $16 million. This had the added benefit for Hoover of putting him in the national spotlight, the hero of the flood relief effort.

*

Back in Greenville, Mississippi, white families had been moved into the second stories of flooded businesses and hotels, while a "colored camp" was set up atop the levees, where African American victims continued to reside. The camp was guarded by armed, white National Guardsmen. The thirteen thousand occupants were required to wear large numbers on their clothing to make it easier to keep track of them. They had to work in order to earn their food, and numerous instances were reported of men being beaten for asking to rest. The supplies intended for all the fifty thousand people affected in Washington County were shipped to Greenville and unloaded by black workers, but they were not the recipients. Separating the races meant that better provisions and medical care could be assigned to white refugees. One African American was shot for attempting to bring food into the camp.

These abuses were horrific, but they were far from the worst or only problems for African American refugees in relocation camps. On May 8, the *Chicago Defender,* the nation's largest newspaper for African American readers, published an exposé of the reality of the Greenville camp. It described "refugees herded like cattle to stop escape from peonage." The *Chicago Tribune* picked up the story, pushing the Red Cross for comments. Prominent progressives, including Jane Addams, the suffragist, social worker, and Nobel Peace Prize recipient, called on Hoover to investigate and stop the mistreatment. The crisis threatened the media image that Hoover had been cultivating.

Hoover responded by asking Robert Moton, the president of the historically black university the Tuskegee Institute in Alabama, to form a Red Cross Colored Advisory Commission to learn whether

African American victims of the flood had been abused "in matter of treatment, living conditions, work details and relief given." This committee brought to Hoover and the Red Cross their draft report on June 14, 1927, documenting cruelties in some of the camps, especially Greenville, where the refugees were forced to work in essential slavery, often beaten or raped by white guards. They confirmed that most of the donations of food never reached the colored camp. In submitting his report, Moton told Hoover to "feel free to make any changes or additions that may seem desirable to you."

The report that Hoover released downplayed any significant problems. It documented minor misdeeds but otherwise praised the efforts of the American Red Cross to help the colored race. Privately, Hoover accompanied the report with promises of reform for the African American community, should he be elected president the following year. He assured Moton unprecedented access to the White House and hinted that he would break up the plantations of bankrupted farmers, allowing African American farmers to own the land they vacated.

Sensing an opportunity, Moton worked in 1928 to support Hoover's candidacy. Southern African Americans were unable to vote in general elections, but they had significant influence in the primaries since, comparatively, very few southerners belonged to the Republican Party. With the help of Moton and Tuskegee, the image of "The Great Humanitarian," swiftly addressing the greatest natural disaster in U.S. history, swept Herbert Hoover first to the Republican nomination and then to a landslide in the general election. Like de Carvalho in Lisbon 170 years earlier, Herbert Hoover discovered the political rewards showered upon a politician who helmed an effective (or seemingly effective) emergency response.

Also like de Carvalho, Hoover seized the opportunity to bring about long-term improvements, politically and structurally. The tension that arose from the unprecedented need of hundreds of thousands who had lost everything; the failure of government-run levees to stop the flooding; and the obvious inadequacies of private donations in relief efforts sparked vehement debate about the federal government's role in disaster management. Every Hearst newspaper carried an editorial demanding that Congress take action. When President Coolidge refused to do so, *The New York Times* for its part praised him for his restraint.

Still, there was a strong push to provide help to the people who had lost so much. Congress moved to create a massive aid bill, one that President Coolidge vehemently opposed. In addition to thinking it outside the bounds of the federal government, he worried that much of the aid had become pork barrel politics, benefiting wealthy southern landowning interests and not the people truly in need. But caught up in the debate was the federal role in flood control more broadly. It was obvious that a very different, much more comprehensive approach was needed if the nation were to have any hope of preventing similar floods in the future.

Hoover used his momentum to create one of the greatest civil engineering projects in history, the 1928 Flood Control Act. It provided congressmen a way of showing that they were acting while allowing the traditionalists to assert that they were not providing individual handouts. With it, the federal government committed to building a massive flood control system for the Mississippi River. It buried once and for all the Army Corps of Engineers' levee-only policy, building reservoirs to keep water out of the rivers and spillways—to redirect it before it could break further levees. The money spent by the affected states in responding to the 1927 flood was counted as matching funds so the federal government could supply the whole cost of the project without setting a precedent for always doing so in the future. The law also exempted the federal government from liability in the case of the system's future failures. Where the act failed, however, was in supporting individual victims. Two-thirds were African American, and their countless, legitimate claims found no real purchase in Washington.

*

The Flood Control Act has guided federal action and massive federal investments in the development of the Mississippi Valley, and the direction of American society, in the century since it was enacted. The lower Mississippi has never again suffered from riverine flooding as it did in 1927. Flooding in 2011 approached the size of 1927, but the effective use of spillways protected the levees. The act also set a precedent for federal dollars being spent on regional infrastructure for the greater good of society—an idea that would fully come into its own under Franklin Delano Roosevelt and his New Deal acts such as the Tennessee Valley Authority and the Works Progress Administration.

Roosevelt, too, owes his election in part to the consequences of the Great Flood. The Colored Advisory Commission's report, as it was released in whitewashed form, may have kept the abuses of refugees from the larger white community, but the African American community was not so easily hoodwinked. Beginning with its coverage of the camps in the spring of 1927, the *Chicago Defender* continued to dig, keeping the issue in front of its readership. It published accounts from refugees throughout that summer, documenting conditions approaching a new slavery. At first, readers appealed to Secretary Hoover, assuming that his Colored Commission had simply missed these cruelties. As time went on, it became clear that he was deliberately avoiding them. In October 1927, the *Defender* published an open letter from the camps by a Mrs. Willis Jones, which resonated with readers.

> We didn't know the Red Cross was supposed to help us till by chance we saw a *Chicago Defender*. We were shocked to see that money and clothes were collected for our benefit while mothers and children lay on straw and naked floors. The most unkind words we received were when we asked for clothes and food from the Red Cross.

Editorials were written decrying the Uncle Toms abasing themselves to the white leadership while their race was reenslaved. The National Association for the Advancement of Colored People, which took a less accommodating approach than Tuskegee, kept the issue alive.

Still, Hoover won most of the African American vote in 1928. How could African Americans vote for anyone *but* the party of Lincoln? The divorce was beginning, though. Hoover had lost 15 percent of the African American vote in 1928—the first time the Republican nominee had received anything less than near-total support.

After the election, it became clear, as it too often does, that the promises that had been made were empty. Moton's calls were ignored, and no discussion of redistributing land in the Delta ever came about. So confident was Hoover that African Americans would never leave the Republican Party that he reneged on everything he had promised or implied.

Hoover underestimated the outrage his betrayal would inspire. By 1932, many in the African American community had decided that Roosevelt's populist, big-government message was a better bet than the duplicity of his Republican rival. Roosevelt won only a third of the African American vote in 1932, but he won 70 percent in 1936. Since then, a Republican nominee has never again garnered more than 40 percent of the African American vote.

*

Natural disasters are disruptions to human systems. Human systems function physically—through sewers and power grids, roads and bridges, dams and levees—as well as socially—through families and friends, churches and synagogues, city councils and legislatures. All those systems have weaknesses, and an extreme natural event puts stress on them. Failures happen where the system is weakest. In Mississippi, we observed a gross failure of the levees, but perhaps more meaningfully, a failure of our society. The Mississippi floods exposed a fundamental weakness in the American social order, a tendency to minimize, dehumanize, and victimize those viewed as other, especially African Americans. The best investment in a resilient community is to identify such weaknesses and repair them—before the event. Such an approach improves lives for everyone during disasters and in between them.

The brutal, lopsided response to the Mississippi flood of 1927 was not uniquely American, as the Japanese attacks on Koreans in the wake of the Kanto earthquake well demonstrate. The evolution of human history might be seen as a gradual expansion of the concept of personhood, from identifying with one's own tribe, to the development of the concept of a nation, to an expanding acceptance of the larger world. You don't have to look far into these examples, or into today's news, to recognize that we still have a long way to go.

Disasters can, at times, bring out the best in us. When the torrent from Mounds Landing first spread over Greenville and the Delta, people who hadn't been able to escape ahead of the waters were left stranded in trees and atop the ruins of houses. The first boats to come to their aid belonged to bootleggers. Although risking exposure themselves, many spent days hunting for and rescuing survivors. In the maelstrom created by a levee break in Arkansas at the height of the crisis on April 20, a steamboat crashed and capsized. An African American named Sam Tucker jumped alone into a rowboat and headed toward the break, where he was able to pull two men to safety.

After the adrenaline of the disaster has passed and we face the dreariness of loss, despair lurks around the corner. Unable to attribute our misfortune to random chance, we wonder what we did wrong. Homes gone, dependent on the goodwill of strangers, fearing financial ruin, perhaps with loved ones killed, we look for someone to blame, we turn on the outsider. A disaster can alter the behavior of the individual, like one who is part of a mob, divorcing us from our moral compass. We must remember the most dangerous threat in a disaster is the threat to our humanity.

CHAPTER SEVEN

CELESTIAL DISHARMONY

Tangshan, China, 1976

When the high and low are in disharmony, Yin and Yang
become confused and bizarre occurrences are born.

—Dong Zhongshu, ca. 150 BC

My first view of Beijing came one week after my twenty-fourth
birthday, in February 1979. I stepped off a plane to a color-
less city. The only billboards bore aphorisms in Mao's calligraphy:
"Serve the people," "Oppose revisionism." China was just recover-
ing from the Cultural Revolution; everyone wore blue or gray Mao
suits; a woman with a colored scarf was the height of daring. When
spring came, the trees brought green into the city's sky, but no grass
grew on the ground. Citizens were required to pull it because it
harbored insects.

I arrived that winter day because I had been chosen to be part
of the first academic exchange between America and China since
the Chinese revolution in 1949. I was a graduate student in seismology at the Massachusetts Institute of Technology, but also fluent
in Chinese after a couple of years in Taipei, Taiwan. My research
proposal was to study the 1975 Haicheng earthquake, a magnitude
7.3 event that the Chinese government claimed had been predicted,
saving thousands of lives. I had a second, less official goal, which
was to understand what had happened in the 1976 Tangshan earth-

quake, which had not been predicted and had destroyed hundreds of thousands of lives. What was the reality? Was earthquake prediction really possible?

Seismology had come a long way since its beginnings in Japan. To be able to monitor compliance with the 1963 nuclear test ban treaty, the United States had created the Worldwide Standardized Seismograph Network (WWSSN), a system of 120 seismic stations around the world. The goal was to be sure we could detect any underground nuclear explosion larger than the test ban limit of 150 kilotons, about the same energy as a magnitude 5.5 earthquake. But that also meant we now recorded every magnitude 5.5 or bigger earthquake around the world. And significantly, that data was not classified. Seismology departments could purchase microfilm sets of all of it.

It changed the way we saw the world. We could now see that earthquakes happened in very narrow bands across the planet, and how those bands correlated with the bathymetry (the depth of water) and residual magnetism of the seafloor—data that had come from naval work in World War II. These were the linchpins of the plate tectonics revolution that upended the earth sciences in the 1960s. We learned that the top layer of the earth, called the lithosphere, was broken into big plates, about a dozen around the world. These plates move around at very slow speeds, no more than a few inches per year. Most of the world's earthquakes happened at the edges of the plates, as they scraped past each other.

Except in China. The only plate boundaries near China seemed to be to the east, in subduction zones off Japan, and to the south, where the Indian subcontinent moves north into Asia, pushing up the Himalayas. And yet China is plagued with frequent earthquakes, many of which, because of China's population density, have been some of the deadliest in history. As the plate tectonics model came together, China was the one place that didn't seem to fit. Why did so much of China have earthquakes? Where were the plate boundaries?

The first answer came in a seminal paper in 1975 by a young American seismologist, Peter Molnar, and a French geologist, Paul Tapponnier. Peter was an assistant professor at MIT, and Paul had come to MIT to work with him as a postdoctoral researcher. In their paper, they showed that India's northward trek was not just pushing up the Himalayas, but also pushing China out of the way, squeezing it to the east. In the same way that picking up a heavy rock is harder work than picking up a light one, as mountains grow taller, it takes more and more energy to continue to push them upward. More rock has to be elevated against the force of gravity. At some point in the earth's history, the Himalayas got so high that it took less energy to move the land that composes China to the east than to continue to push the Himalayas up. Long, fast faults formed that pushed the land of Tibet up and east, and pushed China into the Sea of Japan. Most of the resulting earthquakes hit western China, in the highlands of Tibet, Xinjiang, and Chinghai, but they also diffuse eastward into the northern reaches.

I met Peter and Paul just after their paper came out, as I was applying to graduate schools. Peter had been part of a study team looking into the 1975 Haicheng earthquake, and he was eager to understand whether it had really been predicted, and if so, how. When he received my application to MIT, with my degree in Chinese and physics, he saw an opportunity. Peter told me that if I accepted his offer from MIT, he would do whatever he could to get me to China. I promptly withdrew my applications to other graduate schools.

*

When the Haicheng earthquake struck, on February 4, 1975, China was still embroiled in the Cultural Revolution. China had been through tumultuous times since the success of the Communist Revolution in 1949. The People's Republic of China under the Communist Party was not recognized by most of the world's governments, and at first China relied heavily on financial backing from the Soviet Union. Chairman Mao Zedong, having succeeded in driving out General Chiang Kai-shek and the Nationalist Party, became impatient with the slow pace of reform in Chinese society and of being in the shadow of the Soviet Union. He initiated the Great Leap Forward, to show that China had leaped beyond the socialism of the USSR to true communism, in which society would deploy its labor and distribute funding as Karl Marx had proposed in 1875—"From each according to their abilities; to each according to their needs."

The result was catastrophic, killing by starvation more people in two years than any natural disaster described in this book. Forced collectivization of farms took away the incentive for hard work. People were made to *donate* their farm labor and would be fed whether they worked or not. This led to the most lethal famine in human history, killing at least twenty million people, perhaps as many as thirty million. In 1963, more than half the people who died in China were under the age of ten. In the face of such losses, other members of the Communist Party's Central Committee success-fully sidelined Mao to protect the people from his grand visions of pure communism.

The Party continued to promote Mao as "The Great Helmsman," though, because the propaganda around his cult of personality was effective. For two thousand years, China had had an emperor. "The East Is Red" was an unofficial anthem of the revolution, with the words "The East is red; the sun has risen—from China has arisen Mao Zedong. He brings happiness to the People; He is the People's great Guiding Star." Young Chinese citizens, raised on this propa-ganda, were ready to act at Mao's behest. He forged them into a devastating weapon to reclaim his power.

The ten years of Mao's Cultural Revolution, from 1966 to 1976, were for many a reign of terror. Young people turned on their par-ents and teachers, often encouraged to attack them, even physically. They created public "struggle sessions" where the victims would be publicly humiliated, forced to bow to the children of the Red Guard and be beaten with sticks or chains. Much of the Communist Party leadership at all levels was imprisoned and killed. Teenaged Red Guards roamed the country, joining in attacks on locals who were seen as too intellectual, too reserved, too suspicious. You could be attacked for having received a Japanese education in the 1930s, or for having a relative who worked for the Nationalist government decades earlier. Schools and universities were closed; a whole generation of children was for years denied an education. It is hard for those of us who weren't there to comprehend the level of disruption to all aspects of life this represented. Where I worked, at the Chinese Earthquake Authority (*Guojia Dizhen Ju,* translated then as State Seismology Bureau), the youngest Chinese scientists were thirty-six years old—they had finished graduate school by 1966, just before the revolution began.

The ideology of the Cultural Revolution attacked the Party's elite, but it also had a very strong anti-intellectual component. Early in the formation of the People's Republic, eight categories had been declared enemies of the people: landlords, rich peasants, counter-revolutionaries, "bad elements" (a general category of the criminally inclined), rightists, traitors, foreign agents, and "capitalist roaders" (those who believed in capitalism). With his Cultural Revolution, Mao added intellectuals as a ninth category. They were the "Stinking Ninth" (a pun on a word that could mean either "stinking" or "arrogant"). Many researchers, scientists, and teachers were humiliated, beaten, and even killed for the crime of intellectual work.

The horrors of the Cultural Revolution became the seed of the Chinese earthquake prediction program. In March 1966, as the Cultural Revolution was just starting to take shape, a series of earthquakes occurred near Xingtai in Hebei Province. The series began with a magnitude 6.8 event, followed by several other magnitude 6 earthquakes and finally a magnitude 7.2 three weeks later. Together they killed 8,064 people by official reports and caused damage as far as Beijing, about two hundred miles away. Premier Zhou Enlai went to the epicentral area and urged earth scientists to develop earthquake predictions that could spare future Chinese casualties.

The West had shown no real interest in the seemingly intractable problem of earthquake prediction—Dr. Charles Richter of Caltech was famously quoted as saying that all earthquake predictors were either charlatans or fools. But a prominent Chinese geologist, Dr. Li Siguang, who had led an effort to identify Chinese petroleum deposits and thus develop a measure of energy independence, put his name behind the effort, and it quickly moved forward.

It is impossible to know all Premier Zhou's motivations for initiating such an effort. Many of the scientists I interacted with saw it as a clever move on Zhou's part, a way to insulate at least some of the country's intellectual resources from the worst ravages of the Cultural Revolution. Perhaps Zhou had the foresight to see how attacks on intellectuals could damage China's future. By creating such a program for earthquake prediction, he had a reason to keep scientists working, sparing them from Mao's reeducation camps. One scientist I worked with told me, when we were on a train and couldn't be overheard, that until 1966, he studied Mesozoic structural geology (the geologic history of 65–225 million years ago), but with the Cultural Revolution, he found himself "suddenly fascinated by earthquakes." It was the only safe path.

Intellectuals in a variety of fields were protected through this program. Chinese historians and traditional literature scholars pored over a cache of historical records to create a catalog of earthquakes going back almost four thousand years. It is a priceless scientific resource, the longest earthquake catalog in the world, drawn from the meticulous records kept by the mandarins, the bureaucrats of the Chinese Empire. Even though the researchers' work was sanctioned, they were sent to Tibet to complete it. A seismologist who was part of this team told me it was so that they could keep a low profile during the political turmoil—Zhou's protection could go only so far. She said her father had been a political prisoner, and she volunteered to join the team because she was afraid that her family history would make her a target.

The Chinese scientists developing their prediction program faced the same fundamental problem we Westerners did: the sheer randomness of the earthquake process. There were simply no good theoretical models on which to base our predictions. But with no alternative but reeducation, the scientists tried everything they could think of. Measurements of the earthquakes themselves were made using a network of seismographs, recorded on paper, and read each day. Other instruments were developed and deployed to measure tilt in the ground level, telluric currents (electric currents in the ground), and changes to the chemistry and turbidity of groundwater.

The scientists had to navigate the political hurricane swirling around them. To demonstrate that they weren't part of the "Stinking Ninth," they instituted a program of data collection in what we would today call citizen science. Peasants on farms were asked to report anomalous phenomena—specifically, changes to groundwater such as rising or falling wells, cloudiness, or unusual smells. They noted uncommon animal behavior. The scientists also instituted a program to educate people about earthquakes and their natural causes. The prevailing superstition that earthquakes were caused by problems of yin-yang balance in the government, they felt, was preventing China from addressing the problem.

These programs coincided with a seismically active time. The Xingtai earthquakes of 1966 turned out to be the first of a series in northeastern China. The magnitude 6.3 Hejian earthquake in 1967 and the magnitude 7.4 Bohai earthquake in 1969 suggested a northeastward migration of earthquakes toward Manchuria (comprising the three provinces of China next to Korea). While most big earthquakes don't come in these sorts of clusters, other historic clusters, such as those seen in Turkey in the 1930s and '40s, have led to further damaging quakes, and the series made the scientists nervous.

When, in 1971, the worst of the disruption of the Cultural Revo-

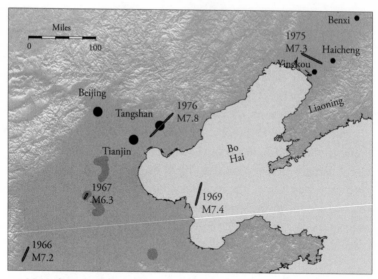

Map of northeastern China showing the causative faults of significant
earthquakes, 1966–76

lution had died down, the Chinese Earthquake Administration was
formed. Three research institutes (in geology, geophysics, and biol-
ogy) were founded in Beijing, and a Seismology Bureau for each
province was established. The Earthquake Administration convened annual conferences, bringing the Beijing researchers and
representatives from the provinces together to discuss where they
anticipated areas of concern for the next year.

Manchuria, especially Liaoning Province, was always on the list,
since the migration of big earthquakes was about the only concrete
phenomenon they had to work with, and those three provinces
were in its path. Instruments were installed in Liaoning, and in
1974, officials began monitoring ground tilt and electric currents.
They found unusual signals. For instance, that summer they found
what appeared to be strong tilting of the ground in the same direction at several sites. This was reported at the annual conference as
cause for concern. The same signal showed up each successive sum-

mer. The scientists had neither the background nor the necessary control readings to rely on, so it was only later that they recognized their data was the result of water being pumped out of the ground for irrigation, rather than data that could contribute to anticipating any earthquake.

Coming into the winter of 1974–75, the scientists in the Seismology Bureau were on edge. They were recording data they had no context for interpreting, aware of the migration of earthquakes but with no basis for guessing which place would be next—and deathly afraid of the consequences of missing a big earthquake. In December, a swarm of small earthquakes started near the town of Benxi in Liaoning, culminating in a magnitude 5.2 earthquake on December 22. That size earthquake was not too common, and it caused great concern. In the next two weeks, various predictions were issued by the local earthquake office for several different locations, but mostly around the area where the magnitude 5.2 had occurred. In some locations, people slept outside for a few days for fear of collapsing buildings. As the swarm of earthquakes died off, the alerts were withdrawn. Weekly reports continued to stream in, often registering strange animal behavior. But the peak of such activity, on Saturday afternoons, seemed to coincide with political meetings exhorting workers to report such anomalies, which were regularly held on Saturday mornings.

On February 1, 1975, another swarm of small earthquakes began. When on the morning of February 4 more than five hundred earthquakes occurred in twelve hours near Haicheng, including a magnitude 4.7 that caused some damage, chaos ensued. Many opted to self-evacuate rather than await word from the government (which itself depended on the slow communication systems of 1975 rural China). The local observatory in Shipengyu that was recording these presumed foreshocks called their local town leaders and told them to expect a large mainshock that night. The movie operator for the community decided to show movies outdoors throughout the night, to convince people to leave their homes. An official in that county, which was named Yingkou, had been extremely active in earthquake preparation, and he called for a formal evacuation.

When the magnitude 7.3 Haicheng earthquake finally came on the evening of February 4, lives were saved as a result of some of these exercises. The capital city of Yingkou county had seventy-two thousand residents, and although two-thirds of the buildings in the town collapsed, only twenty-one people died. The earthquake hit during the movie showings in Shipengyu, and all the people watching the movies were spared. A musical performance had been planned for a visiting military leader, but they canceled the performance and emptied the theater just minutes before the earthquake struck.

The evacuation had been uneven, though, and nothing was done at the provincial level. In the neighboring county of Haicheng, without active evacuations, many more died. The final count of victims showed that for every one thousand collapsed rooms, there were thirty casualties in Haicheng county, but only eleven in Yingkou.

Where there were lives saved, there was political hay waiting to be made. Chairman Mao's nephew, Mao Yuanxin, was a high-ranking member of the Liaoning Provincial Revolutionary Committee. His province, rather than the individual counties that initiated evacuations, was credited with the success. That even the natural world bent to the will of the Party was too good a narrative to be ignored by the leftists, who were struggling to maintain power as the Cultural Revolution dragged on year after year, denying citizens school, safety, and stability. To a scared, resentful populace, the success of Chinese scientists in predicting the Haicheng quake, when the rest of the world had thought prediction was impossible, burnished the reputation of the left. It was broadcast across the country. China had solved the problem of earthquake prediction.

The scientists, of course, knew otherwise. They had been lucky, having observed so many foreshocks. There was no guarantee they could replicate their success.

*

A year later, on July 27, 1976, a magnitude 7.8 earthquake struck Tangshan. A city of 1.5 million people, it was dedicated to coal—its mines were the city's largest employer and so critical to the nation's industrial interests that they operated in three shifts, twenty-four hours a day. Because no big faults had been recognized in the area, Tangshan was considered to have a low level of seismic risk. It had been built with essentially no attention to seismic resistance.

It turned out that a fault ran right under the middle of the city. It was not that large a fault, with a relatively subtle surface expression, and the city had been built before any geologist had looked at the area. In the chaos of the twentieth century, China hadn't had the time or focus to do a systematic investigation of the nation's seismic potential, and many faults lay unrecognized. And just as in Tokyo in 1923, the fault lying directly under the city meant the strongest shaking took place precisely where the greatest concentration of buildings was.

The region was hopelessly unprepared. Almost all homes in Tangshan were either old brick houses or cheaply built multistory apartments. To make matters much worse, the earthquake happened in the early morning, when everyone but those working the night shift in the mines was asleep in their vulnerable homes.

A friend of my family was from Tangshan originally but was living in Hong Kong in 1976. Her extended family, including her mother and five brothers, remained in the city. They lived in some of the new apartment buildings, ten stories tall. The day before the earthquake, her mother had fallen ill, so she went to an infirmary on the first floor of their building. Her illness made her restless. She was awake when the shaking began at 3:42 a.m. She ran to the door to escape, but it was jammed, so she scrambled out a window. From outside, she watched as all ten stories collapsed, killing almost her entire family. Two elementary-school-age granddaughters woke as the earthquake caused their building to collapse, their seventh-floor apartment plummeting to the ground. The rubble trapped them in their bed, but they had been taught in school how to protect their heads and create air space around them as debris settled. They suffered broken bones but miraculously were dug out alive after two days. Every other member of their family who was in Tangshan that night was dead.

It is impossible to tabulate the full extent of the damage caused by the earthquake that tore through Tangshan. In the first months, reports circulated that half of the city, or three-quarters of a million people, had died. The Hebei Provincial Revolutionary Committee originally reported 655,000 dead. By the early 1980s, the official death toll had been revised downward to 242,000. The actual death toll will likely never be known. When I was in China in 1979, the city was still completely closed to foreigners, and those who had worked there told me that only two buildings in the entire city had withstood the earthquake.

With nearly every building destroyed and countless lives lost, normal life simply failed to resume. For days, survivors struggled to dig out other survivors. Beijing knew there had been an earthquake—it is only a hundred miles away, and Beijing had suffered damage itself. But the government was in turmoil. Mao Zedong was near death, and transportation and communication had been disrupted, so it took days to mobilize a response. Getting food and water to the victims proved difficult. Already living in bare subsistence, residents starved.

One group of people who fared better than others were the miners working the night shift. Parts of the mines were flooded, because the movement on the fault changed the flow patterns of groundwater. But no tunnels collapsed and no miners died there. At first glance this would seem surprising, but in fact damage to tunnels in earthquakes is extremely rare. This is true for a couple of reasons. First, the amplitude of seismic shaking underground is only half of what it is at the surface. When a seismic wave hits the surface of the earth, it is reflected back downward, and that reflected wave also causes shaking. So, at the surface, movement is double what you'd find within the earth. Second, tunnels generally have a round or oval cross section, which is a very stable shape.

Large aftershocks continued to rock the region. The city of Tianjin, home to ten million people, was only sixty miles away and suffered one of the worst ones. The city government demanded that the Earthquake Administration send a seismologist to predict further aftershocks. Experts knew the task was scientifically and politically fraught; anyone who could found a way not to be chosen.

The seismologist left with the shortest straw was the most junior researcher at the Institute of Geophysics, the last one hired out of college before the Cultural Revolution closed universities. I'll call her Lao Zhang. She gave daily and weekly reports on the aftershocks and the expected rate of future aftershocks, monitoring their decay. Finally, almost a year after the mainshock, she told the government that she didn't think that there would be any more magnitude 6 or larger events. She was told that "I think" was unacceptable. She had to say yes or no. She chose no. When she told me this story in 1979, she said she had lived the next two years in fear that she would be proved wrong.

Fellow scientists at the Earthquake Administration made no bones about it: Tangshan hadn't been predicted. There had been no precursors. At the annual prediction meeting in early 1976, the Chinese scientists had argued over whether the occurrence of the Haicheng earthquake reduced or increased the risk of another big earthquake in the region. A migration pattern increases the risk of an earthquake until the pattern is over—but how do you know when that is? In the end, they included the northeast on their list of possible sites for an earthquake that year, but did not pinpoint any particular place as most likely.

A geologist with whom I worked—the one who left Mesozoic

structure for earthquake studies—told me a story about Tangshan. He said that the day before the earthquake, the Hebei Provincial Seismological Bureau received a report that some wells near Tangshan were behaving anomalously. Some had turned artesian—the water level had increased to the point that water flowed out of them like a spring. This can occur naturally for a variety of reasons. But two scientists from the bureau were going to be traveling past Tangshan anyway, so they were asked to stop in and investigate. They got into Tangshan late at night and went to sleep at a guest hostel, planning to look into it the next day. The hostel collapsed in the earthquake, and both scientists were killed. I asked him how often wells were reported turning artesian without an earthquake. He said it was quite common.

*

It was in the two months after the earthquake that the political and scientific worlds really began to merge. Even more than in Japan, Chinese governance was rooted in the philosophy of Dong Zhongshu, the second-century-BC pedant who had fused Confucian politics with yin-yang mysticism. For two millennia, aspiring scholars had had to pass exams on the classics, including the works of Dong Zhongshu, to enter government. The Mandate of Heaven justified the emperor's rule, and a natural disaster was a sign that the mandate had been withdrawn. Dong's admonition to the emperor to issue an edict of self-criticism after a natural disaster was still standard practice up until the end of the empire in 1911. The imperial order and the nation's scholarly bureaucrats may have been swept away by the Communists, but a deeply ingrained mythology doesn't change overnight.

Mao Zedong had explicitly cultivated the adulation of the public, and he filled the cultural niche once inhabited by the emperor. When the Tangshan earthquake hit, Chairman Mao was dying. Medical doctors had moved into the Forbidden City, and Mao had not been seen since April. Significantly, he did not go to Tangshan in the aftermath of the earthquake. Rumors of his illness were already swirling through the country. They merged with the prevailing superstitions about earthquakes. The classical texts were clear on the subject: an emperor's death (or impending death) would trigger an earthquake.

In August 1976, a newspaper in Taiwan made the connection explicit, calling the Tangshan earthquake an omen of Mao's impending demise. Between the public misgivings about earthquake prediction and the possibility that one earthquake wasn't enough to restore the yin-yang balance, earthquake panic swept the country. Half of the provincial seismology bureaus issued predictions, and, as was seen before Haicheng, even more people decided on their own to stay out of buildings. My colleagues said that in August 1976 as many as five hundred million people may have been sleeping outside. A magnitude 7.2 occurred on August 16 in the mountains of Sichuan Province, when most of the almost one hundred million residents of China's largest province were already living outside. For the tiny fraction of those hundred million who lived in the epicentral area, this practice may have saved lives. But the earthquake further stoked fear and uncertainty.

Mao Zedong died on September 9, setting off a power struggle. Many books have been written parsing the dynamics, in the aftermath of his passing, between the leftists who had carried out the Cultural Revolution, led by Mao's wife Jiang Qing, and the more moderate members of the Party. A month after Mao's death, a coup by the moderates arrested Jiang Qing and three others, labeled the Gang of Four. They moved quickly and effectively, detaining them without bloodshed (despite strong support for the leftists in many provinces and major cities). The response to the Tangshan earthquake could well have been a factor in the quick dispatching of the leftists.

Tangshan fit with several of Dong Zhongshu's other warnings.

In addition to a dying emperor, the two other main causes of excess yin were said to be ministers having usurped the power of the emperor and the presence of women in government. Both charges were laid against the Gang of Four. In the indictments, the leftists were accused of using Mao's name and prestige to accomplish their own ends. And Jiang Qing's indictment emphasized her womanly wiles, the flagrant sexuality of her relationship with Mao. This of course played into deep-seated cultural prejudices about women, ones beyond any relationship with natural disasters. But it didn't hurt, in their prosecution, to have such a manifest example of the consequences of a yin imbalance.

No one ever officially blamed the Gang of Four for causing the Tangshan earthquake. The Communist Party was working hard to eliminate these types of superstitions. But the Party used phrases from Dong Zhongshu's treatise in the indictment to subtly make the tie back to the earthquake. The undercurrent in public thought is evident. Big earthquakes usually cause rumors of more earthquakes to come in the area affected by the event—as they should, given the prevalence of aftershocks. However, the mass self-evacuation of half a billion people, as was seen in China in 1976, was unprecedented. Their actions, across the country and so far from Tangshan, were driven by traditional superstition.

When I was in Beijing in 1979, the Gang of Four was in jail but not yet on trial. I was a novelty, one of only thirty-five Americans in Beijing and the first scientist. I wasn't completely free to talk with the Chinese, but there were many situations—in the residence compound, restaurants, taxis—where my job as a seismologist was discussed. I was astonished at the widespread belief in the government's ability to predict earthquakes. One car driver proudly told me that the Chinese didn't have to worry about earthquakes anymore because the Party would give them warning. What about Tangshan, then? I asked. He, like the majority of nonscientists I spoke to, assured me that Tangshan had been predicted, but that the Gang of Four wouldn't let the seismologists share it. He told me that they didn't want anyone talking about an earthquake that they had caused by taking over the government, by usurping Mao's power.

*

I came home from China with a changed view of earthquake prediction. I completed a study of the physics of the Haicheng foreshocks, showing that they had actually delayed the mainshock. (That raised our hopes that we could find a discriminant for foreshocks—something that made them look different from other earthquakes—but we never did.) I also came to understand that earthquake prediction was, at its heart, not a scientific problem. Or at least not *just* a scientific problem.

When I discuss the inherent randomness of the timing of disasters, there is one obvious exception: the ability of one earthquake to trigger another. Omori first quantified it over a hundred years ago. But it turns out the earth doesn't have its own hard definition of what an aftershock should be. Overwhelmingly, aftershocks are smaller than mainshocks. But at the outer extremes of the probability distribution, occurring 5 percent of the time, are aftershocks that are *bigger*. Similarly, while most happen close in time and space to the mainshock, we sometimes see longer clusters that include many large earthquakes, like those we saw in northeast China.

When I went to Beijing, these ideas were still developing. Chinese scientists didn't know any more about prediction than we did. But unlike us, they lived in a political climate that forced them to act on their guesses, something we American scientists could never do. In rural China, where bad construction increased the benefit of anticipating an earthquake, and where a farming economy decreased the costs of a false alarm, guessing could be worthwhile. In America, where many more people die in traffic accidents than in earthquakes, the economic costs of evacuation could have been devastating, and the public perception, distilled in the free press, presented a huge political downside to false alarms. The same information is less actionable.

I spent the next decade of my career trying to quantify the probabilities that one earthquake would trigger another. I thought that if we did our part to collect and interpret seismologic information and then translated it into probabilities, we could hand it over to policy makers and emergency managers, who would synthesize it with social, political, and cultural considerations, and decide what course of action needed to be taken. It was naïve of me, I realize now, to think that anyone can make decisions on the basis of probabilities alone. It was a lesson that I put to use much more effectively twenty-five years later.

CHAPTER EIGHT

DISASTERS WITHOUT BORDERS

The Indian Ocean, 2004

[E]very person is a new door, opening up into other worlds.

—John Guare, *Six Degrees of Separation*

One of the most important tenets of research science is the acknowledgment that the easiest person to fool is yourself. All human beings, including scientists, are inclined to confirmation biases—we are less critical of information that supports our pre-existing beliefs and more critical of data that confounds them. The scientific process, especially peer review, was developed to help researchers recognize when we are not seeing the data clearly. In peer review, we take our research—our intellectual offspring, the prized result of our hard work—and put it in the hands of a colleague, or even a competitor, and ask him or her to tear it apart, to find the flaws and tell us what we did wrong. It can be so emotionally difficult that many new PhDs will decide they'd rather pursue a career that doesn't involve research. This is an unfortunate downside of a process that is critical to finding out what is undeniably true.

One consequence of having our work regularly torn to shreds is that scientists tend to be very careful to say just what we mean. We never present a result before laying out all the background and describing the full experiment. We avoid using vague qualifiers. If

I were to say, "This is a *big* earthquake," a colleague would be ready to tell me that, in fact, it is not as big as another earthquake. Or that it is "big" only if you neglect site amplification, or don't take into account the paleoseismic record. Or, or, or . . . We can say "big" only after defining precisely what "big" means.

With each disaster, scientists are asked, "Was this the Big One?" (Those implied capital letters always seem to be part of the question.) To answer—to analyze and compare disasters—we need an approach that lets us quantify the variations between them. So, in each field, scientists have come up with a scheme for classifying the relative size of an event, each based on some measurable, indisputable quantity. Maximum wind speed is used to classify hurricanes—the Saffir-Simpson scale—and tornadoes—the Fujita. Volcanoes have a Volcanic Explosivity Index (VEI), defined by how much material they've thrown off, to what height, and for what duration. (It's a 0–8 scale. Vesuvius in AD 79 was a 5; Laki in 1783 was a 6.) Floods are the only natural hazard classified by their probability of occurrence; a hundred-year flood has a 1-in-100 chance of happening in a given year. For earthquakes, seismologists created magnitude, representing the total energy released in an event.

Each scale represents a physical measurement, something that can be defended to colleagues, reduced to a simple number that seems easier to explain to the public. But none of them reflect the damage experienced by individuals. Destruction is difficult to quantify; terror is impossible to measure. Scientists are much more comfortable living in the defined, physical, quantifiable world.

And these physical measurements are not *wrong*. They do what they are supposed to do—define what happened physically in the world. The problem is that when someone asks, "Was this the Big One?" there is a disconnect between the language used by the person asking, in human terms, and the scientist answering, in terms of physical effect.

Sometimes, however, the physical and the human are in sync— the Big One is truly *the Big One*.

The magnitude 9.1 earthquake and tsunami that hit the west coast of Sumatra, Indonesia, on December 26, 2004, is just such an event. The physical scale of it was unprecedented. The length of the fault that moved in that earthquake was over 900 miles—the longest earthquake rupture ever seen. (The earthquake that leveled San Francisco in 1906 was on a fault 275 miles long.) It took fully nine minutes for the rupture front to move from one end to the other.

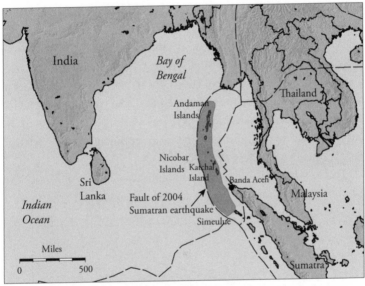

The fault surface that moved in the Sumatran earthquake, 900 miles long and almost 100 miles wide, is shown in the hatched area.

The earthquake took place in a subduction zone, called the Sunda Arc. In purely physical terms, subduction zones give us the world's biggest earthquakes. We've seen how the size of an earthquake is dependent on the distance of the slip and the length of the fault on which the earthquake occurs. Yet there's a bit more complexity to it. Plates can be as much as fifty miles thick. The deepest parts are so hot that friction can't hold them in place; the rock becomes ductile, able to deform itself, pulling like taffy instead of breaking. It's at shallower depths—the first ten to fifteen miles—that friction holds the edges of plates immobile. Here the rock bends, storing elastic energy. When friction is overcome and one side slides suddenly past the other, we observe an earthquake. Its magnitude, then, is primarily determined by three factors: the distance of the slip, the length of the fault, and the depth at which friction is held. The San Francisco earthquake of 1906 involved a comparatively small depth of eight miles. But the fault it was on was almost three hundred miles, and the slip was significant enough that, taken together, they created a magnitude 7.8 earthquake.

We've also seen earthquakes in subduction zones, in which one plate is pressed beneath another, like a truck pushing up and over a small car in a rear-end collision. These plates generally meet at a shallow dip, just five to twenty degrees. The rocks being pushed under are colder, because they have only recently been close to the surface. Cold temperature means more opportunities for friction. This results in a much *wider* fault.

To see what this all means, imagine two faults, each two hundred miles long, one vertical and one subhorizontal. The vertical

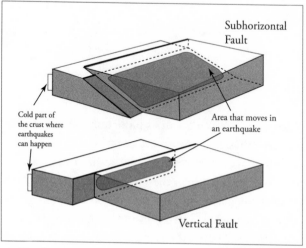

How heating of the crust controls the width of earthquakes

fault has earthquakes in just its top ten miles. So the *area* of the fault could be said to be two thousand square miles. A subhorizontal fault, in a subduction zone, can have earthquakes in its top twenty miles (because the rock being pressed down is still relatively cold and not yet ductile). But with a ten-degree dip, there are 115 miles of fault that lie in that twenty-mile band. The area of that two-hundred-mile-long fault is now twenty-three thousand square miles. The same length of fault has an area more than ten times greater. Add to this the fact that subduction earthquakes tend to have greater slip, and the result is a much bigger earthquake.

Only the biggest subduction zone earthquakes can produce ocean-spanning tsunamis. A magnitude 8.0 can produce a tsunami that is devastating locally, but it will not have displaced enough water to create waves that can traverse ocean basins. If, however, you hear that there has been an earthquake with a magnitude of about 8.5 or larger, you can be pretty sure that it is on a subduction zone and that a tsunami will immediately follow.

A tsunami is a wave with multiple peaks and troughs. Instead of thinking of it as a very big ocean wave, you need to think of it more like the ripples that encircle a rock that you drop into a pool of water. A series of up-and-down movements travel out from a disturbance of the ocean floor. How many peaks and troughs, their relative sizes, and how far apart they are all depend on the shape of the seafloor, not to mention the shape of the coastline where the wave hits. If a trough arrives first, then the first sign of an incoming tsunami is not rising but receding seawater.

If a tsunami wave is twenty feet high, then everything that is less than twenty feet above sea level goes underwater. If there is a thirty-foot cliff at the ocean's edge, the tsunami will not reach land. If the sea level is already two feet higher than average because it's high tide, the tsunami wave will be two feet higher, submerging anything below twenty-two feet of elevation. The damage is usually done not by flooding, but by the currents. The wave will first travel across the ocean at the speed of a jet plane, but as the seafloor shallows, it will slow to the speed of a car. Still, a huge mass of water moving at twenty miles per hour has an almost incomprehensible amount of energy; anything not very securely anchored in place will be carried away. Cars and people are easily swept off. Weak buildings can be pushed over. Strong buildings can have all the walls stripped off them, leaving just the frame. I've seen pictures of an apartment in a strong building where the only item left was the refrigerator. It was too heavy to be carried away, left in its place draped in seaweed.

The Sumatran earthquake had the third-largest magnitude ever recorded. It represented the longest fault we have ever seen slip. It displaced the third-largest volume of water ever recorded and produced by far the biggest wave we have ever seen in the Indian Ocean. The energy released in the event was a thousand times larger than that of the largest hydrogen bomb ever detonated. Its human impact was no less momentous.

*

The 2004 magnitude 9.1 earthquake itself, for all its movement, did not affect as many people as its size might suggest. The fault largely ran under the ocean and sparsely inhabited islands. The earthquake began just before 8 a.m. the day after Christmas. Aceh Province, at the north end of the island of Sumatra, was at the southern end of the earthquake rupture, and its proximity to the fault meant it suffered some of the worst effects. Many buildings in the province of Aceh and its capital, Banda Aceh, were badly damaged from the shaking. Tsunami waves arrived soon after the mainshock ended, even as aftershocks continued. Along the western coast of Aceh, wave height was fifty to one hundred feet. The number of dead is uncertain, because many people were washed out to sea, their bodies never recovered. We know that the total number of dead or missing in Indonesia was more than two hundred thousand, three times as many as in any other country, and most were from Aceh Province. The first floors of most buildings in Banda Aceh, a city with almost three hundred thousand residents, were flooded, and

10 percent of the population perished. Leupeung, a town of ten thousand also on the west coast of Aceh, was completely obliterated; only a few hundred survived.

The small islands north and west of Aceh sit on the subduction zone and so were both shaken strongly in the earthquake and hit quickly by the tsunami. Due to the contours of the seafloor and the coastline geometry, some islands suffered much larger waves than others. The northernmost of the islands saw tsunami heights of only five to ten feet, while Katchal Island in the Nicobar Island group was hit with a thirty-five-foot tsunami that killed 90 percent of its population. The indigenous tribes on Katchal lost essentially all their tribal leadership, their way of life, and their culture. On another island, Simeulue, the community's memory of a tsunami in 1907 spurred them to head to higher ground as soon as the earthquake was over, and very few died.

The next country hit after Indonesia was Sri Lanka, a thousand miles west of the fault. Earthquakes produce waves of many frequencies, and just as with sound, the lower-pitched waves travel farther than the higher-pitched ones (in the same way that, when hearing music from far away, you might hear the lower frequencies of the drumbeat but not the higher frequencies of the melody). Consequently, although in a nearby earthquake you will feel a high-frequency, jerky motion, distant earthquakes are often experienced as a slow rolling, which some reported feeling in Sri Lanka. The tsunami wave took about ninety minutes to traverse the Bay of Bengal. It wrapped around this island country, hitting all the coastlines at heights from thirteen to forty feet. Because many of the towns were near the coast, their wooden buildings were easily carried away, and more than forty thousand people died.

Tsunamis produce their strongest waves directly perpendicular to a fault. Since the fault in the Sumatra earthquake ran approximately north–south, its impact was felt gravely to the west in Sri Lanka, but also in Thailand to the east. The west coast of Thailand is a popular beach destination for tourists from across the world, and its hotels were full for the Christmas holidays. When the waves arrived, two hours after the earthquake, they were second in height only to those hitting Indonesia, with some reaching sixty-four feet.

The tsunami continued its inexorable progression across the Indian Ocean, taking lives in India, Malaysia, the Maldives, and Myanmar. It reached the east coast of Africa, killing people in Yemen, Seychelles, South Africa, and Kenya. The waves spilled out of the Indian Ocean into both the Atlantic and Pacific Oceans, their continued movement recorded for the next few days on monitoring instruments of the National Oceanic and Atmospheric Administration. In total, it struck thirteen countries, causing damage to infrastructure and buildings in five more. Forty-seven others lost citizens who were traveling abroad, many of them vacationers in Thailand. In this respect, it was not only a massive physical event with a devastating human toll; the Sumatran tsunami was the world's first truly global disaster.

*

In 1977, Kerry Sieh got his PhD in geology from Stanford University with research so innovative that he was immediately offered a professorship at Caltech. Eventually, his work led to the creation of a whole new field of research called paleoseismology. Not bad for a twenty-six-year-old. Up until that point, geologists had studied faults by examining the features and layers of rock that were offset in earthquakes, and by assessing how the surface of the earth had been changed as a result (called geomorphology). Kerry's innovation was to recognize that there was much more information available, right at the fault's surface, if we could somehow get in and see it. So he came up with the idea of digging a trench across a fault and mapping out, inch by inch, the features that had been moved by it. He used a site in a marsh where new ground is rapidly created as the dirt settles in stagnant water. These layers can be dated by measuring the carbon-14 in the organic material. When the San Andreas moves in an earthquake, it cuts through all the layers; but after the earthquake, new layers will be laid down on top. By mapping out the layers that were cut (or not cut), Kerry could create a history of earthquakes—on a fault that had moved just once in our recorded history.

From that beginning, Kerry and his students, and later their students, continued to develop the field, digging into faults across California and beyond. This work allowed us to know that on average every one hundred to two hundred years, all of the San Andreas moves in a big earthquake—information that would have been impossible to glean from the historic record alone.

Subduction zones, however, with their faults offshore, pose a unique problem, their pasts hidden. We could tell from the fast motion of the Indian Ocean plate that earthquakes in the Sunda Arc had to have been pretty common, but our written history wasn't even one hundred years old. The 1907 earthquake was well enough remembered by the elders of Simeulue to spare that tribe, but it formed the extent of their—and our—historic record. That was until Kerry came up with another idea.

Porite corals grow along the Sunda Arc, and they thrive in the sunshine—to a degree. They grow, adding rings to themselves annually, until they've reached just below the waterline and no farther. After that, their growth runs sideways. If the ground were to move up and push them above the surface, the exposed part of them would die. And if the ground sank, they would resume their vertical growth until they again reached the water's surface.

Along a subduction zone, plate motion holds one slab down, allowing the corals above it to grow to full height. But when the slip of an earthquake happens, a plate snaps up, pushing the coral above sea level and killing the parts that have been exposed. Could this phenomenon be used in some way to draw a seismic history from subduction faults?

In early 2004, Kerry and one of his students, Danny Natawidjaja, published a major paper documenting great earthquakes on the Sunda Arc that had occurred in 1797 and 1833. This meant that

Coral that died because of its emergence in the 2004 tsunami.
From John Galetzka, Caltech Techonics Observatory

instead of knowing about one earthquake (in 1907), we now know that there have been three earthquakes in the last 250 years. This makes it easier for us to stipulate that earthquakes happen pretty often in that location, about once per century, give or take. And that means that another was in our future, most likely in the lifetime of many people alive now.

Danny was Indonesian, and that summer he was heading back to Indonesia to finish the last round of fieldwork for his PhD. All too aware of the risk posed by tsunamis, and knowing how unprepared his country was, Danny and Kerry created posters explaining what a tsunami was and how moving away from the coast after a big earthquake could save lives. The posters, written in English (for tourists), Indonesian, and the regional language, Mentawai, were distributed and posted in their field area in the summer of 2004, several months before the Sumatran earthquake. It is difficult to quantify harm that didn't happen, but there are undoubtedly those who survived the tsunami because of Danny and Kerry's efforts. Not many scientists get to see the results of their work so clearly.

It begs a bigger question, though: Why were so many lost when a warning could easily have saved them? I received an email fifteen minutes after the 2004 event began, explaining that a magnitude 8.8 (estimated) earthquake had occurred off the north end of Sumatra. I get an email like this for every magnitude-5-or-greater earthquake in the world—and so can anyone else who wants to; the USGS Earthquake Program website offers such a service. So I knew almost immediately that a deadly tsunami was likely striking the Indian Ocean. I sat in front of my Christmas tree in California, knowing a catastrophe was unfolding but powerless to do anything about it.

Getting wide-scale warnings to people at risk is possible, but it requires infrastructure: you need the instruments to record an event, people to man a warning center, and the means to convey an alert to those at risk, as well as their governments. In 1946, an earthquake in Alaska created a tsunami that killed 150 people in Hawaii, leading the U.S. government to create the Pacific Tsunami Warning Center (PTWC) in Hawaii in 1949. After the deadly 1960 Chilean earthquake led to a tsunami that killed dozens in Hawaii and hundreds more in Japan, the warning center expanded into an international operation, providing alerts for the entire Pacific Basin. When the 2004 Sumatran earthquake hit, the PTWC issued a statement that there was no tsunami risk to the Pacific Basin—in itself true. But it was made in line with the limitations of the PTWC, which permitted warnings only for the western side of the Pacific Ocean.

The last time an earthquake that large had registered was forty years earlier, with the 1964 Alaskan earthquake. But the technology available at that time could not discriminate the true size of our largest earthquakes. It was only with the advent of digital recording and processing in the 1980s and '90s that we could even see the extra energy contained within the earth's biggest earthquakes. The PTWC personnel still relied on older technology, so they initially understated the size of the event. By an hour later they'd realized the earthquake was bigger than they had anticipated, but there were no established channels for reaching the governments in the affected countries. Even when they did get through, those countries didn't have mechanisms for conveying warnings to people on the coasts. It is hard for scientists to contemplate how many lives might have been spared if our data had been better communicated.

Fortunately, steps have since been taken. We saw with the California floods of the 1860s how we tend to minimize threats that have faded from our collective memory. The flip side of that tendency is to respond aggressively to those perils that are still fresh in our minds. The enormity of the losses suffered in Sumatra, combined with the knowledge that a technological solution could have prevented deaths, has led to action. Less than two weeks after the tsunami, a proposal was put together to implement a warning system for the Indian Ocean and to address some of the antiquated technology. A system is now in place, coordinated by the United Nations, using data systems from Australia, Indonesia, and India. The Pacific Ocean warning systems have started using more modern techniques for estimating size, leaving the world less vulnerable to future disasters.

*

Most of us have heard of the notion of six degrees of separation, the theory that any person on the planet is connected to any other person on the planet through a chain of acquaintances with no more than five intermediaries. I would guess that the chain between anyone in the world and a victim of the Sumatran tsunami rarely exceeds three. (My brother had a colleague at work who never returned from his Christmas holiday in Thailand. Los Angeles has a large Sri Lankan community; I am sure one of the Sri Lankans I know knows someone who was killed.)

Fifty-seven different countries lost citizens, almost one-third of the world's countries. Some countries lost more citizens in the tsunami than from events at home. The Sumatran tsunami killed more Swedes than any other natural disaster in their history, more than any single event in their history since a battle in 1709. Globalization and the convenience of air travel have changed the world in fundamental ways. For the first time, many if not most people throughout the world shared in the loss from this one natural disaster.

Advances in telecommunications also fundamentally deepened the impact of this disaster. Pictures of the devastation traveled around the world faster than the tsunami itself. On our TVs and computers, we saw images of flattened houses, water surges, massive ships tossed to shore as if they were toy boats. Then when your neighbor's cousin or your child's teacher's nephew didn't come home from vacation, it became more than a remote event; it was a horror that had extended into your own life.

This heightened awareness of disasters on the other side of the world is changing how we look at and deal with disasters. Part of what inhibits us from taking action to prevent human catastrophe is our uncertainty about *when* an event will happen. A risk that has a lower probability of occurring in any given year is inevitably displaced by more immediate concerns. By definition, the Big One, one that can destroy a society, is rare. The annual or decadal flood gets factored into our urban planning, and seismically active areas adopt building codes to withstand typical earthquakes. But most of those more frequent disasters are part of a continuum that has, at its end, a much rarer, much worse event.

What constitutes a very rare event locally is, in global terms, much more common. California has a magnitude 8 earthquake only every few centuries, but somewhere in the world, a magnitude 8 happens almost every year. In the past, we didn't have the means to see globally. At the time of Pompeii, a Roman citizen had no concept that Indonesia existed, let alone that a volcanic eruption there had destroyed a human settlement. When Laki erupted in 1783, a few scientists in Europe realized that something was going on in Iceland, but most were oblivious (it was almost a year before its own government in Denmark sent aid). When the Kanto earthquake destroyed Tokyo in 1923, the news was transmitted to the United States by telegraph, but Americans had no way of understanding the hell that had been endured. Even when the Tangshan earthquake killed two or three times as many people as the Sumatran tsunami, just forty years ago, the world noticed, but generally with little regard. Not only were almost all the victims Chinese, at a time when China was isolated from the world, but we didn't witness the pain. This was before the Internet or personal computers brought the world to our desks. Even pictures were withheld—China didn't want the world to see its weakness. No foreigners were even allowed into Tangshan for five years after the earthquake.

In that light, if the Sumatran tsunami could offer us anything other than grief, it was awareness. Knowledge of tsunamis has never been more prevalent. We still struggle to raise awareness among the right people—too many on coasts are still vulnerable, too many who live above five hundred feet elevation fear for themselves unreasonably. The role of subduction earthquakes in shaping the greatest tsunamis, allowing us to predict where they'll strike, is not well enough understood. But the word *tsunami* means considerably more to us than it did even twenty years ago.

The decade since the Sumatran tsunami has seen substantially more interest in coping with natural disasters. The UN has established the Office for Disaster Risk Reduction, whose "Sendai Framework," negotiated at a meeting in Sendai, Japan, in 2015, has been adopted by the General Assembly. Globalization and modern telecommunications for the first time turned a local disaster into an international experience.

Humanity's expansion of its definition of "us," from our family to our tribe to our nation, is continuing its outward course. In the Sumatran tsunami, we saw an "us" that encompasses the world, and in the process we changed the way we look at disasters. It caused the world to experience a catastrophe in the kind of visceral way that impels us to overcome our most threatening, deeply ingrained biases.

CHAPTER NINE

A STUDY IN FAILURE

New Orleans, Louisiana, United States, 2005

Anything coming up this street darker than
a brown paper bag is getting shot.

—A white resident of Algiers Point in New Orleans

*T*here, but for the grace of God, go I. A phrase we use to express
identification with a victim. A Los Angeles–area news radio sta-
tion ran a special just after Hurricane Katrina struck New Orleans,
in 2005, called "There, but for the Grace of God," making explicit
the connection between the storm and California's own suscepti-
bility to catastrophe. At its best, the phrase demonstrates a recognition of our common vulnerability, a sympathy for the suffering.
For many it operates, too, like a kind of talisman, a shield against
the randomness of disaster. If I trust enough in God's goodness, I
will be spared the same fate. But we often become less charitable in
speculating *why* God's grace was withheld from the victims.

We've looked at the human tendency to seek out patterns in
disaster (even when such patterns prove specious). It's a behavior
that has, over the millennia, proven lifesaving—when we made
the connection, say, between someone's violent gastrointestinal
distress and the mushrooms they ate. But the handmaid to causation is *blame*. When we hear that a person has had a heart attack,
how quickly do we leap to considering her lifestyle, her weight?

When we're told someone has been diagnosed with cancer, we often ask, "Did he smoke?" Consciously or not, by assigning a person the blame for his or her own misfortune, we are inoculating ourselves from the same fate. *I am active,* we might silently assure ourselves. *I don't smoke.*

It is, in part, our impulse to find blame that made the idea of natural disasters as divine retribution appealing. Think of the burghers of the Netherlands refusing to offer aid to the victims of the Lisbon earthquake of the eighteenth century—that it was not their place to undo the punishment God had decreed. In this view, their Calvinist faith offered them protection against the punishments being visited upon the Catholic idolaters. While the development, and now the prevalence, of scientific models for natural disasters has largely taken away from us these simple explanations, it hasn't diminished our need to insulate ourselves, by assigning blame to the victim.

This is nowhere truer in modern American history than with Hurricane Katrina. Katrina was America's first catastrophic natural disaster since the advent of television. It killed more Americans than any other event since the 1906 earthquake in San Francisco, and it came close to destroying an iconic American city. The images of New Orleans horrified us. We saw our fellow citizens abandoned to the rising waters, dying, standing helpless on their rooftops. We saw them herded like animals into the Superdome, the city's football stadium—left without power or light, left to defecate in hallways. It was an experience most Americans would have considered unthinkable in our country.

And as reports of the catastrophe washed over the rest of us, our minds attempted to answer the inevitable, impossible question: *Why?*

*

Of the meteorological disasters, tropical cyclones are the deadliest. Depending on where they form, they can take different names—sometimes hurricanes or typhoons—but all represent the same basic phenomena: rapidly rotating weather systems, characterized by strong winds and a spiral arrangement of thunderstorms. We refer to them as hurricanes when they form around North America, in the Atlantic Ocean or the eastern Pacific Ocean.

All storms require a source of energy to keep their water in the air, and the air moving. For tropical cyclones, the source is the air found just over the ocean near the equator. Because this water is warm, so is the air above it, which rises, carrying moisture upward. This leaves less air near the surface, creating areas of lower pressure. This mechanism—warm air rising, lower pressure near the surface—is what sustains a hurricane, and why hurricane season peaks at the end of summer. A hurricane can form only when the water temperature of the top 150 feet of the ocean is at least 80°F, and that is most likely to occur after months of long days heating the ocean.

Of course, hot air rises everywhere—there need to be more conditions in place than just a warm ocean for a hurricane to form. First, the area of higher temperature must be surrounded by cooler areas. When the hot air rises and lowers the pressure of a region, the air around it, at higher pressure, will flow into the low-pressure area. That "new" air becomes warm and moist, and it too rises, perpetuating a cycle.

As the water vapor rises into the higher levels of the atmosphere, it approaches cooler air. The differential between the warmer air and the cooler air causes the vapor to condense back into water droplets, forming clouds. The energy that was needed to evaporate the water gets released in the process. The air is now even hotter, causing it to rise even more.

This process gets water into the air, but the storm's rotation—the high winds characteristic of a hurricane—is dependent on the Coriolis force caused by the earth's rotation. The Coriolis force is zero at the equator and increases toward the poles. Hurricanes can form only in the band that is far enough away from the equator (at least three hundred miles) to get spin going, but close enough to the equator that the water temperature is at least 80°. The storm's rotation pulls even more air into the region of lower pressure.

The last factor in the formation of a hurricane is the absence of what's called vertical wind shear. That means the direction and speed of the overall wind pattern can't change much as air rises up through the atmosphere. If the rising hot air were to hit winds blowing in different directions, it wouldn't continue its straight ascent—it would get pulled sideways, disrupting the storm's formation. It is only when all these factors are in place that a hurricane can take shape.

Because hurricanes are driven by warm ocean temperatures, most scientists expect to see an increase in both their number and strength as global warming progresses. Indeed, the strongest hurricane ever recorded, as defined by peak wind speed, was Hurricane Patricia in the eastern Pacific in 2015. Hurricane Harvey in 2017 carried more rain to the Houston, Texas, area than ever before seen in one storm. Hurricane Irma the same year maintained extreme winds for longer than ever recorded. Still, the year with the most named storms (those that reached at least tropical storm strength with winds of at least thirty-nine miles per hour), and the most hurricanes in the Atlantic Basin, was 2005. (The intense hurricane season of 2017 had almost as many major hurricanes as 2005 [six in 2017 compared to seven in 2005], but fewer smaller hurricanes [ten as compared to fifteen in 2005].) And of all the storms in 2005, none inflicted more harm than Hurricane Katrina.

*

America's response to natural disasters had come a long way since President Coolidge had, after the 1927 floods, rejected the idea of providing direct payments to citizens. The widespread suffering produced by that disaster triggered a public outcry. It led to the enactment of the Flood Control Act of 1928 and a substantial federal investment in flood management of not just the Mississippi River, but several other major rivers across the nation. Even though

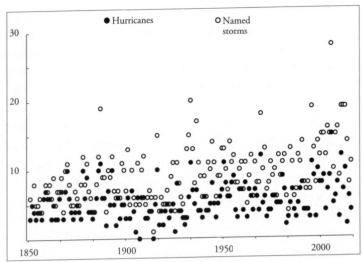

Graph depicting Atlantic Basin storms from 1850 to 2015. Data from the National Oceanic and Atmospheric Administration.

that expenditure did not benefit individual victims, it set a precedent for federal government involvement in the largest of natural disasters.

Soon after the Mississippi floods, a combination of farming practices and wide-scale drought led to the ecological and social disaster that came to be known as the Dust Bowl. This was during the Great Depression, and after the election of Franklin Roosevelt, it prompted a more proactive government response. Roosevelt's administration created several agencies aimed at helping displaced farmers, as well as avoiding the practices that contributed to the Dust Bowl in the first place. This further cemented the outcome of the 1927 floods and the government's role not just in helping victims, but in devising longer-term mitigation strategies.

The next few decades brought further federal disaster relief, one disaster at a time, until 1950, when Congress passed the Federal Disaster Relief Act. For the first time, Congress authorized the expenditure of federal funds toward disaster recovery (finally putting to rest Grover Cleveland's pronouncement that "people support the government; the government does not support the people"). However, programs were developed by different agencies as needed. In the 1970s, disaster relief was provided in some instances by more than one hundred different governmental agencies, creating confusion and inefficiency.

It wasn't until 1979 that the Federal Emergency Management Agency was established to consolidate the response. And because the primary function of FEMA was to distribute money after a disaster, it became a home for political appointees. Giving out government money, even in the context of a disaster, always offers political advantages. In 1992, FEMA had the distinction of having the highest ratio of political appointees to career civil servants of any agency in the U.S. government.

The 1990s saw a change in the attitude of and toward FEMA. When President Bill Clinton appointed James Lee Witt, former head of the Arkansas Office of Emergency Management, as the organization's head, it was the first time the head of FEMA, or any high-level federal political appointee, had had any experience in emergency management. Witt understood the political value of responding well to natural disasters, and his response to Mississippi floods in 1993 and California earthquakes in 1994 demonstrated a competence that was a political asset to the Clinton administration. Witt also understood the value not just of relief but prevention, establishing several programs to reduce losses from disaster, such as buying people out of their property on floodplains and retrofitting buildings to withstand earthquakes and severe winds.

One enduring aspect of Witt's tenure at FEMA was planning—specifically, designing scenarios (along the lines of my efforts with ShakeOut) to both anticipate likely disasters and prepare the government's response. Regional offices of FEMA developed catastrophe plans relevant to their regions.

In Louisiana, FEMA developed a scenario for what would happen if a Category 3 hurricane hit New Orleans and breached the levees, causing widespread flooding. They called it Hurricane Pam.

*

The Mississippi River at New Orleans is an *active delta*—the mouth of a river large enough that the settling of sediment, combined with rising sea levels, creates a migrating system of distributaries, or smaller branches that flow away from the river's main stream. This means that the location of "the river" is not stationary; the arrangement of outflowing branches changes from season to season. There are only about seventy active deltas around the world, including those of the Nile and the Ganges. New Orleans, however, has the distinction of being the only major city situated *within* an active delta.

Sediment and sea level interact in a complex system. When a river hits the sea, the speed of its current slows to close to zero, depositing at the mouth of the river the sediment it had been carrying. Meanwhile, rising sea levels (ascending for the last thirteen thousand years, since the end of the last ice age) cause sediment to drop *earlier*. Together these forces raise the level of a riverbed. We've seen the way levees naturally form adjacent to rivers—and also how a flood will inevitably break through them, carrying water to lower elevations. This is what leads to the creation of new distributaries, but it has another effect as well. The weight from the sediment these floodwaters deposit will also, over time, cause the crust beneath the flooding to sink a bit. It creates a subsidence—an indentation—in which more sediment can settle.

The massive flood control projects enacted on the Mississippi since 1927 have changed the dynamics of the river. It now carries less sediment than it used to, much of it caught in reservoirs upstream. Still, the weight of the remaining sediment, confined within levees, is significant; the crust beneath the delta continues to subside. The result of this interplay is that the Mississippi is rising while the land around it is settling. Much of New Orleans now sits below sea level, some of it as much as twenty feet below.

The levees were built to protect New Orleans, but they are man-made systems, some old and insufficient to handle what Katrina threw at them. This was known before 2005. Three years prior, the Louisiana Water Resources Research Institute at Louisiana State University had completed a scientific study showing how the process of delta formation had turned New Orleans into a deep bowl waiting to be filled by a storm surge. The confinement and destruction of wetlands along the coast compounded the problem. Researchers predicted that in a slow-moving storm with high rainfall, storm surge and rain runoff would *overtop*, exceeding the height of many of the levees.

Their study formed the scientific basis for the Hurricane Pam scenario. In it, five exercise days were planned, to test what would be needed in the event of such a hurricane—factors including search and rescue, evacuation procedures, and emergency supply maintenance. Four of these exercise days had been completed when Hurricane Katrina turned Hurricane Pam into reality.

Physically, the actual storm was very close to the hypothetical one—the scenario predicted to within 10 percent the total rainfall and level of flooding. Many of the social and engineering consequences were accurately predicted as well, including the number of people evacuated and resettled in public shelters, boat-based rescues, chemical plants affected, amount of debris, destroyed buildings, and collapsed bridges. In the weeks after Katrina, Secretary of Homeland Security Michael Chertoff said, "That 'perfect storm' of a combination of catastrophes exceeded the foresight of the planners, and maybe anybody's foresight." But it simply wasn't true. Emergency management professionals knew exactly what would happen to New Orleans, and his own agency had already been planning for precisely the storm that they faced.

*

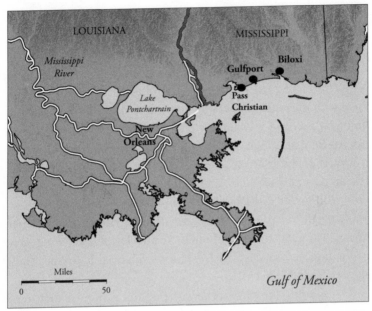

Map of the delta of the Mississippi River

We know Hurricane Katrina as a storm that hit New Orleans, but in fact, it devastated a large part of the Gulf Coast. Katrina formed near the Bahamas, strengthening to hurricane status just before it crossed Florida on August 25. It weakened as it passed over land, but once it got to the Gulf of Mexico, it recovered its strength and more. It made landfall in the early morning of Monday, August 29, with its eye, the circular center of a hurricane, passing east of New Orleans and into Mississippi.

Hurricanes are categorized by wind speed. We call them tropical storms when they exceed 39 miles per hour. Category 1 begins at 74 miles per hour. Category 5 hurricanes are those in which maximum wind speed exceeds 157 miles per hour. But wind speed tells only part of the story. Hurricanes inflict damage in three ways: (1) wind tears structures apart; (2) water is pushed onto shore in what's called storm surge; and (3) the rains themselves cause flooding. Wind is a factor in the first two, but not in the third, and thus a slow-moving Category 1 can cause more rain, and potentially more damage, than a fast-moving Category 4.

The fastest speeds of a hurricane are near the center of the storm, with a bit more energy in the quadrant northeast of the eye. With this quadrant hitting the state of Mississippi, the level of destruction near its coast was almost total. The storm surge that hit was twenty-eight feet high, destroying essentially everything within a half mile of the coast and reaching as far as twelve miles inland. The larger cities of Biloxi and Gulfport saw widespread destruction, with gutted buildings and casino barges washed far onshore. Many small towns were wiped out. Pass Christian, one of these towns, had eight thousand homes, and all but five hundred were badly damaged or destroyed. The total financial losses in Mississippi exceeded $125 billion.

Considering the physical devastation, it's a wonder that more lives weren't lost. The three coastal counties had a combined population of four hundred thousand people. But the state of Mississippi began issuing evacuation orders on Saturday, August 27, and the region had been largely vacated by the time the worst of the hurricane struck. The total death toll across Mississippi was 238.

While this number is significant, the fact that it wasn't much greater is a tribute to storm prediction. We've considered the limits of prediction when it comes to earthquakes, which solid-earth scientists (such as myself) can offer much insight into but which are still *random about a rate*. Atmospheric scientists can do a much better job of it—they have the benefit of seeing what they are studying. Successful prediction in any discipline requires evidence of an event preceding the occurrence you're trying to predict. For earthquakes, while stress must first accumulate in the earth, it is difficult to observe it through miles of rock, nor can we tell how it accumulates differently for large earthquakes compared to small ones.

Hurricanes, on the other hand, cannot strike out of the blue.

First a storm has to form over the ocean, build up energy, and travel to land. All this happens in the atmosphere, where it can be observed—from satellites and through airborne measurements. The challenge then is not in determining whether a hurricane is coming, but rather in predicting the strengthening or weakening of the storm, as well as its path. Data collection systems of the last couple of decades, combined with comprehensive modeling made possible by supercomputers, have resolved these two questions with often astonishing accuracy. The National Weather Service (NWS) short-term predictions for Katrina nailed the storm track within fifteen miles and the wind speed within ten miles per hour.

On Friday night, August 26, fifty-six hours before Katrina's eventual second landfall, the NWS issued an ominous warning to residents of the Gulf Coast.

MOST OF THE AREA WILL BE UNINHABITABLE FOR WEEKS . . . PERHAPS LONGER . . . HUMAN SUFFERING INCREDIBLE BY MODERN STANDARDS.

Not all states responded as quickly as Mississippi, with its mandatory evacuations. The state of Louisiana and the city of New Orleans both waited until nineteen hours before landfall to mandate evacuation, leaving residents with precious little time to respond.

Over the next two days, the scenario that had been described in Hurricane Pam played out in real time. Although the storm was less intense in New Orleans than it had been in Mississippi, the city still suffered immeasurable damage as Hurricane Katrina passed to its east on the morning of August 29. High-rise buildings suffered blown-out windows. Sections of the roof of the Superdome were stripped away. Those who remained on the coast felt the storm's impact firsthand. In the first few hours after landfall, the Coast Guard rescued sixty-five hundred people from trees and rooftops.

Conditions only worsened. Just as the LSU Louisiana Water

Resources Research Institute report predicted, the storm surge, extreme rain, and high winds were more than the city's levee system could handle. That Monday several overtopped; others breached. (Hurricane Pam had foreseen the overtopping. It hadn't predicted breaches.) The first was reported to the NWS on Monday morning, soon after landfall. On Tuesday, more breaches occurred. Compounding the problem, many pumping stations—which would have otherwise removed water from the city—stopped working due to power outages and flooded equipment. By Wednesday, 80 percent of New Orleans was underwater by as much as twenty feet.

This was the physical damage. The impact to human systems was no less devastating. Sewer, drainage, power, supply chain, and communication systems all went down. To those remaining in New Orleans, everything that we have come to expect of life in America had disappeared. Many found themselves with no choice but to travel to the city's designated refuge of last resort.

Close to ten thousand people rode out the storm at the Superdome, and after the levees failed, thousands more poured into the already overcrowded stadium. Supplies were inadequate for the numbers and the duration of the emergency. On Tuesday morning, the U.S. Department of Health and Human Services assessed the Superdome—no lights, no air-conditioning, no functioning sewage system. They deemed it uninhabitable. Maybe so, but almost twenty thousand people were inhabiting it.

The result was a hell most Americans could hardly fathom. The *Los Angeles Times* quoted Taffany Smith, a twenty-five-year-old at the Superdome with her three-week-old son. "We pee on the floor. We are like animals," she said. The horrors were unceasing. "At least two people, including a child, have been raped," the *Times* reported. "At least three people have died, including one man who jumped 50 feet to his death, saying he had nothing left to live for."

The situation was hardly better for those who stayed in their homes. Residents were forced into their attics and onto their roofs.

Others drowned in their own houses. In the end, the Coast Guard alone rescued thirty-three thousand people, and tens of thousands more were rescued by other agencies, as well as by their neighbors. We don't know how many people died in New Orleans from Hurricane Katrina. Louisiana reported 1,464 victims one year after the hurricane but by its own admission never finished counting them. Resources had to be devoted to more pressing concerns.

*

These two images—the Superdome and its savage conditions, and families waving to passing helicopters for rescue from their rooftops—became the enduring visual symbols of Hurricane Katrina. These were what the rest of the world took in from their televisions, as they witnessed the devastation visited upon the Gulf Coast. Among the emotions they elicited was sympathy. The American Red Cross received almost $1 billion in donations for hurricane relief in the first month alone.

Katrina also presented an object lesson in the way that, in trying to find a pattern in the disaster that would let us believe we wouldn't suffer the same fate, we unfailingly look for someone to blame. It offered no shortage of scapegoats, but two narratives, not necessarily mutually exclusive, dominated the discussion—the failure of the government and the failure of the victims.

There is plenty of evidence to support the idea that the government failed to ensure public safety, its fundamental obligation. In a bipartisan committee report issued in 2006, Congress called the response to Katrina "a failure of government, a failure of initiative, a failure of leadership." Failure of this magnitude requires failure at all levels of government; it is not nearly so simple as saying that FEMA was to blame. The American system of emergency management is based on the premise that all disasters are local. Response begins with local officials, who have complete authority to respond. If they are overwhelmed, they call for help from the state, handing over both authority and responsibility. If the state is overwhelmed, it can call for help from FEMA. But FEMA's role is primarily as a dispenser of money. Much of what went wrong with Katrina wasn't under its purview.

The reality is that the government failed its citizens before, during, and after the event. The levees were ill suited to the flooding they would face. Built by the Army Corps of Engineers, they had been turned over to the New Orleans Levee Board for maintenance and were to be inspected by the Army Corps every year. Analysis after Katrina showed that their design was known to be inadequate; the Levee Board had ignored their requirement to be trained in levee maintenance; and annual inspections tended to be social affairs rather than thorough examinations. Millions of dollars in Levee Board funds were spent on repairing a fountain in a park while they ignored a floodgate that had been damaged in a train accident and could not be closed.

The emergency planning for hurricanes was close to useless. Hurricane Pam offered a remarkably accurate picture of what was to come, but the city planned for much less. Army Lieutenant General Russel Honoré, who arrived in New Orleans on Wednesday, August 31, to lead the military's support operations, described the situation at the Superdome as "a classic example of officials thinking about the worst-case scenario but providing only enough resources for the best-case scenario." Supplies were inadequate. There was no emergency operations center. The city didn't know how the national incident command system worked. Where Mississippi began evacuations fifty-six hours before landfall, Governor Kathleen Blanco of Louisiana and Mayor Ray Nagin of New Orleans delayed evacuations. Problems with communicating these orders weren't taken into account. Search-and-rescue teams weren't supplied with boats. The list is endless.

One of the greatest failures came in the lack of cooperation between levels of government. For instance, the state of Louisiana asked the state of California, through the Emergency Management Assistance Compact—a mutual aid agreement between states that enables them to share resources during natural and man-made disasters—to send a team of experts to help reassemble the New Orleans city government. Los Angeles sent fifteen people—experts in search and rescue, law enforcement, and city services, led by Fire Chief Daryl Osby (now fire chief of Los Angeles County). After a briefing with Governor Blanco in Baton Rouge, he and his team arrived in New Orleans to discover that Mayor Nagin had no idea he was coming. Chief Osby later said, "It wasn't until I got there that I understood the dynamic, between the federal, state and local governments. How they weren't communicating. It was finger-pointing."

In the response, both the New Orleans and Louisiana governments made mistakes that arose from not understanding how the emergency management system functioned. Mayor Nagin was operating out of a hotel room, because the city did not have an emergency operations center. Chief Osby and assistants from other states helped the city assemble an operations center over the next two weeks. Governor Blanco did not understand how the process for coordinating federal and state resources worked, relying on ad hoc tutorials from her National Guard.

Corruption further impeded response and recovery. Chief Osby described his first meeting with Mayor Nagin: "Mayor Nagin said to me, 'Thank you. But I really don't need you. If you can just help me have FEMA write me a check for $100 million then we can handle this.' I had to ask him to repeat that. I explained to him there really is a process." In the weeks that Chief Osby was there, he was regularly offered payment to direct response resources a certain way.

More than two hundred police officers, 15 percent of the force, failed to respond after the hurricane passed. Some of them were dealing with their own legitimate family crises. Others, it turned out, hadn't returned to service. After the publicity surrounding the hurricane, fifty-one were fired for desertion. A Department of Justice investigation into the New Orleans Police Department after Katrina concluded that the NOPD was dysfunctional at all levels: "We found deficiencies in a wide swath of City and NOPD systems and operations." They included misuse of force, lax hiring and supervision, and corruption.

The mobilization of federal resources to support recovery provided a great new vista for corruption. Billions of dollars poured into New Orleans, falling into cracks between the fighting city, state, and national governments. The public outcry against FEMA helped ensure that funds streamed in, but much of it got diverted. Mayor Nagin left office in 2010, but he was convicted in 2014 on twenty of twenty-one bribery and tax evasion charges. He has the dubious distinction of being the first New Orleans mayor to be convicted of corruption. The state of Louisiana suffered from corruption as well. A program administered by the state called A Road Home, intended to help people rebuild and return to New Orleans, received $1 billion in federal money. An investigation in 2013 found that 70 percent of this money, or $700 million, was unaccounted for.

*

The government's failure to protect its citizens in the case of Hurricane Katrina was significant, and the impulse to blame them understandable. But to the observer unconsciously struggling with the fear of suffering a similar fate, it offers a less than fully satisfying answer. There is some solace in knowing that governments can be replaced. Indeed, Governor Blanco didn't run for another term after Katrina, facing clear evidence of her unpopularity. But most of us feel disempowered by our governments—and how can we be sure that one failed government isn't followed by another? It can lead us to a way of thinking that's harder to justify but no less insistent.

There are those victims, we might argue, who simply made the wrong choice. The approximately one hundred thousand people who remained in New Orleans defied evacuation orders. Evacuation would have improved the odds of their survival, as the coastal regions of Mississippi demonstrated. *We wouldn't have made the same mistake,* we might tell ourselves.

Of course, the truth is much more complicated. We've seen how little time New Orleanians had between their evacuation order and Katrina's landfall. For many of those left behind, escape was impossible. How were those without cars, one-quarter of the city's population, supposed to escape? City plans had accurately estimated the number that wouldn't have the means to self-evacuate, but no other option had been provided to them. The city didn't use its fleet of school buses to transport them out, later claiming it couldn't take on the liability and that it didn't have enough bus drivers.

If buses *had* been operational and residents had escaped, where would they have gone? Many couldn't afford lodging. The hurricane arrived at the end of the month, and people on limited incomes were waiting for the checks they expected to arrive in two days. For many the Superdome presented not the best but the *only* option.

The media invited in all of us a tendency to find fault in the victims, describing what it observed as rampant lawlessness and rioting. "Looters take advantage of New Orleans mess," declared the Associated Press. "Relief workers confront 'urban warfare,'" said CNN. Lieutenant General Honoré reported that when he arrived in Louisiana on August 31, Governor Blanco was disappointed that he had not brought a larger number of troops to handle what she was sure was the breakdown of civil authority in New Orleans. A similar narrative spread, dominating much of the news coverage. The speed with which this message was disseminated (and the fact that many such reports were later disavowed) speaks to an unsavory appetite awakened by disaster—the need of onlookers to distance themselves from the victims, to subtly ascribe them some measure of blame.

General Honoré heard the descriptions from the media and from many in the government that portrayed a city under siege, but what he witnessed in New Orleans himself was something very different. He saw people in desperate straits, fighting to survive, demonstrating the "patience of the poor." In time, much of the evidence for widespread lawlessness that pervaded media coverage was proven false. An MSNBC report showing New Orleans law enforcement engaged in "looting" was demonstrated many months later to be the police acting at the direction of their supervisors, getting supplies for people in need. Five years after Katrina, the *New York Times* reported that, rather than the narrative put forth of an African American population terrorizing the city, "a clearer picture is emerging, and it is an equally ugly one, including white vigilante violence, police killings, official cover-ups and a suffering population far more brutalized than many were willing to believe."

These reports, unlike those of looting, have withstood the test of time. A woman I know was a teenager in Baton Rouge at the time. She described neighbors running to buy guns when they heard New Orleans evacuees were coming to their town. A mixed-race group of refugees were attempting to escape across a bridge that led to the mostly white city of Gretna. In a chilling echo of the Mississippi floods nearly a century prior, they were confronted by city authorities who shot over their heads, ordering them to turn around, back into the floodwaters.

Vigilante groups formed in mostly white, mostly unflooded Algiers Point, assaulting any African Americans who showed up in their neighborhood. Some of them have been indicted. *The New York Times* reported that one defendant, Roland Bourgeois Jr., was quoted as saying, "Anything coming up this street darker than a brown paper bag is getting shot." His trial, postponed many times, was delayed indefinitely in 2014.

On the Danziger Bridge, two African American families, trying to get away from the chaos of New Orleans, were gunned down without warning by four New Orleans police officers responding to a call of shots fired. Those policemen were convicted; on appeal their sentences were greatly reduced.

*

Everyone who watched the horrors of Katrina in New Orleans came away believing that events didn't have to unfold the way they did. Harder, however, is considering how they might have been handled differently. Failure at such a massive scale requires many smaller failures from many players. But just like a branch breaks off a tree where the wood is already rotten, the biggest failures, with the biggest consequences, occurred where a system—physical, political, or social—was already weak.

The levees of New Orleans failed because they were designed to defend against an encroaching river that was always going to win in the long run. The Mississippi flood control system is one of the most awe-inspiring feats of human engineering, but it is fighting an impossible fight. Effective management of the river requires accepting that it will move and learning to accommodate those changes rather than simply attempting to diminish them. There is a reason that no other major city lies in an active delta.

The government undeniably failed its citizens in the case of Hurricane Katrina, but the failure resulted from dysfunctions that were pervasive long before the hurricane formed. The distrust between city and state made cooperation impossible. The corruption that has long plagued New Orleans bogged down the recovery and kept the citizens there in misery.

We saw the way too many Americans found their African American compatriots to be victims not of circumstance but of their own choices. The tendency, conscious or not, to impute blame to those suffering is a response to natural disaster so common as to feel inevitable. We naturally resist the idea that suffering might be caused by forces outside our control, and so, to reassure ourselves, we assign the responsibility to the sufferer. It is just as human as the impulse toward charity, and just as unlikely to be extinguished. But by being aware of this tendency, we can learn to recognize it around us. When the next disaster strikes, we don't have to fall victim to it ourselves.

CHAPTER TEN

TO COURT DISASTER

L'Aquila, Italy, 2009

Only fools and charlatans predict earthquakes.

—Charles Richter

As a graduate student at MIT, I was once quoted in a newspaper article about earthquake prediction, related to my work in China. Soon after, I received a letter from a man in Scotland telling me that he knew how to predict not only earthquakes, but also "volcanoes, hurricanes, storms, fires, murders, heart attacks, rapes, and other natural disasters." Who saw murder and rape as natural disasters? In four pages of small single-spaced type, he expressed his contorted vision of the world. I read it in shock. A senior seismologist said to me, "Welcome to seismology. Time to start your wacko file."

Any seismologist with a reasonable degree of public exposure receives this sort of letter with depressing regularity. I have received letters and phone calls from people who predict earthquakes with numerology, the phases of the moon, water dowsing (a debunked method for discovering underground reservoirs), creative Scripture reading, and even their own physical ailments. One lady would regularly call the USGS office in Pasadena predicting earthquakes for San Francisco when she had a headache and for Los Angeles when she had diarrhea. (As a fourth-generation Angeleno, I found this rather insulting.) Another would go out each morning and make drawings of the slug trails in her driveway, predicting earthquakes for places whose coastlines resembled them. We received a fax with those drawings almost every day for years.

As we've seen, mankind despises randomness, often resorting to desperate measures to construct predictable patterns. Most people don't resort to throwing a Bible down the stairs and using the page to which it falls open as a basis for their suppositions. But most cultures with earthquakes have, for instance, created a myth of "earthquake weather." My mother lived through the 1933 Long Beach earthquake; for her the foggy weather, common in March when that earthquake occurred, was earthquake weather. To those whose first significant earthquake was the 1987 Whittier Narrows earthquake, it was the hot Santa Ana winds that hit us that day, as they often do in October. After the trauma of the earthquake makes us notice the weather, our need for a pattern, alongside our confirmation bias, impels us to notice the times the pattern seems to fit and ignore the times it doesn't.

But the number of people calling us with their predictions was exceeded by a different constituency: those convinced that we knew precisely when earthquakes would hit but were unwilling to share our information. They would rather believe that I was lying than concede that an earthquake's timing simply could not be predicted. "I know you can't tell me when the next earthquake will be," one woman wrote me, "but will you tell me when your children go to visit out-of-town relatives?"

*

Italians have been trying to predict their earthquakes for millennia. Pliny the Elder, in his *Naturalis Historia,* advanced one of the first written theories of earthquake weather. "I certainly conceive the winds to be the cause of earthquakes; for the earth never trembles except when the sea is quite calm and when the heavens are so tranquil that the birds cannot maintain their flight . . . nor does it happen until after great winds." He also suggested that they were more common in the mountainous regions of Italy: "I have found by my inquiries that the Alps and the Apennines are frequently shaken."

While his temporal predictions have not stood the test of time, his spatial patterns have proven fairly reliable. The Apennines have been one of the most frequent sites of Italian earthquakes. The seismic hazard map of Italy shows a region of highest risk down the spine of the boot. This is the result of the region's complex plate tectonic setting. On the grander scale, as we saw with Pompeii, the African plate is moving north into the Eurasian plate. But it is complicated by smaller pieces, called microplates, which jostle around the boundaries. The Adriatic plate is a microplate underlying the Adriatic Sea, to the east of Italy, and it seems to be moving independently of both the African and Eurasian plates. Specifically, part of the Adriatic plate is subducting under Italy, thus forming the Apennine Mountains that run the vertical length of the country.

More so than in many earthquake-prone regions, the earthquakes in the Apennines tend to come in clusters. We have seen how, in the earliest days of seismology near the turn of the twentieth century, Fusakichi Omori put forth an equation to describe how one earthquake triggers another. While the underlying principle is true for all earthquakes, the parameters of the equation vary between regions, and between individual earthquakes. One magnitude 7, for instance, might trigger a small aftershock sequence, with just one or two magnitude 5 aftershocks (like the 1989 Loma Prieta near San Francisco); another might be followed by hundreds of magnitude 5s. It could be followed by a larger earthquake, like the magnitude 7.2 off the coast of northern Japan in 2011 that was followed two days later by a magnitude 9. While any individual occurrence is hard to anticipate, we do see regional tendencies in these variations. The Apennines are one of the regions where clusters of small earthquakes are common. That said, from time to time those clusters *will* include a larger, damaging earthquake. The clusters can go on for days, weeks, or months before they get bigger—or not.

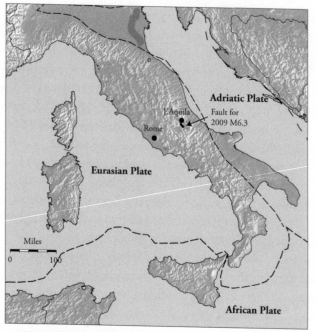

Map of Italy showing plate tectonic boundaries and the L'Aquila earthquake

This type of cluster presents real problems for seismologists trying to talk about risk. A study done by the National Institute of Geophysics and Volcanology (INGV) in Rome found that about 2 percent of these Italian swarms include a large earthquake. That means that if a cluster has begun, there is about a 2 percent chance of a damaging earthquake during its duration. Seen another way, there is a 98 percent chance that there will *not* be a damaging earthquake. Nonetheless, this still represents a significant increase in risk. Consider the difference: In a given location, a damaging earthquake happens only once in hundreds of years. A random month has, say, a 1-in-10,000 chance of seeing a major earthquake. The 1-in-50 chance once a cluster is under way, then, represents a two-hundred-fold increase in risk.

Still, it is only 1 in 50. Forty-nine of those fifty clusters will produce nothing. So what should be said to the public? That the risk has gone up by two hundred times? Or that there is a 98 percent chance that nothing will happen?

*

Seismologists have a love-hate relationship with earthquake prediction. The drive to make patterns and predictions is baked into any scientist's DNA, and yet earthquake prediction has always fallen just tantalizingly out of our grasp. Seismologists began cataloging earthquakes in the earliest twentieth century in the hope of discovering meaningful patterns. Harry Wood, one of the early giants in seismology, wrote a proposal in 1921 to install the first seismographs in Southern California, justifying it by saying that if we could see where the small earthquakes happened, this information might let us know where the big ones would be. This has turned out to be only partly true—some small earthquakes occur near the major faults, but the San Andreas Fault stays quiet, breaking only in the largest earthquakes. And the small earthquakes have not turned out to have any clear information about the *time* of big earthquakes, outside the basic aftershock-triggering pattern described by Omori in 1891. These hopes are weighed against the bad science and even fraud committed in the name of earthquake prediction. That's to say nothing of that constant stream of hopeless letters from would-be predictors. We have learned to be very skeptical of any such claims.

After the first data was collected in the 1920s and '30s and the lack of pattern became apparent, most scientists focused instead on understanding *why* earthquakes happen—at least until the Haicheng earthquake, which, with its unlikely survivors, made it clear *something* had gone right. Formal government programs were re-initiated in the United States, Japan, and the Soviet Union in hopes of solving this riddle once and for all.

The Haicheng earthquake showed us some of the Chinese approaches to prediction. Most of the research there and elsewhere centered on the idea that in order for a major earthquake to happen, stress had to build up along a fault, and therefore we should look for evidence of that increasing stress. Strainmeters, instruments that directly measure the deformation of the ground, were installed along faults. The Chinese investigated shifts in groundwater chemistry. It made a kind of physical sense; if rocks were highly stressed and started to crack, they could release gases into the water, changing its composition in a demonstrable way. The cracking could, arguably, also change the electrical conductivity of rocks in the vicinity. There was even a controlled experiment in California, in line with China's citizen science, to put to the test once and for all whether animals could perceive earthquakes before we do. Farmers along the central part of the San Andreas Fault, where magnitude 5 earthquakes are pretty common, were enrolled in a program to report their animals' behavior. They issued reports at least weekly and were not allowed to submit a report about the behavior of the animals before an earthquake *after* the earthquake took place (to make sure that what they reported wasn't influenced by their knowledge that an earthquake had indeed occurred).

But as the years went on, these investigations mostly came to naught. Strainmeters registered changes, but analysis showed they were reflecting not seismic stress but changes in groundwater levels, as California's boom-and-bust climate fluctuated between drought and flood. Water chemistry studies were equally disappointing. There had been hope that radon gas in groundwater might be a useful indicator. Radon is produced by radioactive decay, especially in granites, and it would be reasonable to infer that increased radon in the water was a result of ground cracking, and thus seismic activity. But a careful study in Iceland convincingly demonstrated that there was no change in radon before volcanic eruptions (which put even more stress on rocks than earthquakes). And, no, animals cannot predict earthquakes. Controlled studies showed no more reports of unusual behavior before earthquakes than at any other time. Smart young researchers concluded that their careers would be better served by investigating topics more likely to be successful, and so after a couple of decades of focus, interest in prediction once again faded away.

The nature of earthquakes is such that you can appear to be successful in finding patterns for quite a long time before you have to face the fundamental failure of your prediction. Earthquakes happen all the time, after all. It is remarkably easy to be right just by random chance. A magnitude 5 earthquake happens somewhere in the world every eight hours. If you declare, "There will be a magnitude 5 earthquake tomorrow" without specifying a location, odds are you'll be right.

Some exploit this fact for fraud. I remember in 1994 hearing about a man who sent a fax to a company in Los Angeles predicting that a magnitude 6-plus earthquake would occur in the next week. When his prediction was borne out with the Northridge earthquake that caused $40 billion in damage, the company was impressed. They were ready to listen to his proposal that they buy further earthquake predictions from him. What they didn't know was that he had sent the same fax to a different company each week, waiting for random chance to make him look successful.

More insidious is our ability to fool ourselves. Imagine you're a scientist who's predicted a magnitude 5 for a particular region at a particular time. There is some random chance that one will happen. Let's say that likelihood was 5 percent. Now a magnitude 4.7 earthquake happens in your window. It seems like you were pretty close, right? You should be able to count that as a success, shouldn't you? But the random chance of a 4.7 in such a scenario would be not 5 percent but 10 percent—twice the likelihood. And you've allowed your definition of what constitutes a successful prediction to be defined by what actually happened, undermining the value of the statistics. Compound this with a little flexibility as to the location and the timing, and even well-intentioned predictions begin to lack meaning.

Many a researcher has fallen down this rabbit hole—until stopped by a statistician. Most seismologists now insist that even an apparently successful prediction could be specious. A method needs to show predictive validity over multiple events and be demonstrably better than random chance. We have fallen victim, over the decades, to too many false alarms to demand less. Otherwise, our confirmation bias will fool us into thinking we've discovered a pattern, when in fact all we've done is create constellations out of randomly arranged stars.

*

In January 2009, a swarm of earthquakes began near the ancient Italian city of L'Aquila. L'Aquila was built in the Middle Ages by Frederick II, the Holy Roman Emperor, as a walled city to protect a federation of ninety-nine villages. He was defending them—and his own realm—against the growing political power of the papacy. Thus its name, which means "the eagle." It has been the transportation, commercial, and communication center of the area for centuries, and it stands now as the capital of the Abruzzo region of Italy. It sits high in the Apennines, at twenty-eight hundred feet of elevation, and more than seventy thousand people call it home. It has a long history of earthquakes—deadly ones have been recorded from 1349 (eight hundred victims), 1703 (three thousand victims), and 1786 (six thousand victims).

The swarm that began in January continued through February and March 2009, with numerous felt earthquakes. In light of the city's seismic history, its residents' nerves started to fray. Schools were evacuated several times in the first few months of 2009.

Enter Giampaolo Giuliani, a resident of L'Aquila and a technician at the National Laboratory of Gran Sasso, part of the National Nuclear Physics Institute. Giuliani worked on machines designed to detect radioactive gases, and by early 2009, he had been examining earthquake patterns and their relationship to the presence of radon for nearly a decade. The new swarm offered an opportunity to put his ideas to the test. In February 2009, he put forth a set of predictions, based on his radon measurements, which were given to the media and inspired a variety of stories. He never submitted a written prediction to authorities, so we don't know exactly *what* he was predicting. We have only the media reports.

In response to these predictions, scientists at the government's earthquake research center, the INGV, made public statements consistent with the current understanding of earthquakes in the region: that these types of swarms were common, reliable earthquake prediction was still impossible, and the risk of big earthquakes was still low. Although these are true statements, they did little to assuage public anxiety. In mid-March, as the earthquakes continued unabated, an Italian blog, *Donne Democratiche,* asked for Giuliani's opinion about the ongoing earthquake activity. He said that the swarm was a "normal phenomenon" for the region, not a precursor to a larger event, and would be diminished by the end of March.

But on March 30, the largest event up to that point—a magnitude 4.1—struck L'Aquila. Giuliani offered another prediction in response. He told the mayor of the town of Sulmona, thirty-five miles southeast of L'Aquila, to expect a damaging earthquake within six to twenty-four hours. The mayor responded. Vans mounted with loudspeakers drove through the town broadcasting warnings and urging residents to flee. Many abandoned their homes, but ultimately no earthquake struck.

So Giuliani had at least twice made definitive predictions that failed to bear out. Meanwhile, earthquakes were being felt regularly, and the media was still turning to him for comment. The Italian government struggled to contain him. Authorities told Giuliani he was needlessly panicking residents; but despite his previous failures, more than a few residents believed his predictions.

The government needed to craft a message as convincing as Giuliani's. It fell to the Italian National Commission for the Forecast and Prevention of Major Risks, the formal government mechanism for communication between the National Civil Protection Service (what in the United States is often called Emergency Services) and the scientific community. The committee, composed of earthquake scientists and engineers, meets once a year in Rome to review research and monitoring activities, but could also be called on an emergency basis to evaluate an imminent risk.

In an unusual move, the government convened a special meeting of the Major Risks Commission in L'Aquila on Saturday, March 30, that lasted just one hour. The only minutes of that meeting were compiled one month later, so they offer a not particularly reliable source. We know that authorities went into the meeting to create a reassuring narrative. The head of Civil Protection, Guido Bertolaso, was recorded saying as much in a pre-meeting wiretap, which had been initiated as part of a different investigation. In particular, he said that the experts were going to tell the public, "It's better that there are 100 magnitude-4 tremors rather than silence, because 100 tremors release energy and there won't ever be the damaging tremor."

After the meeting concluded, the six scientists in attendance left immediately, while the vice director of Civil Protection, Bernardo De Bernardinis, gave a press conference about the meeting. He echoed his boss's earlier wiretapped assertion about the benefits of swarms. "The scientific community tells us there is no danger," he said, "because there is an ongoing discharge of energy. The situation looks favorable." In reply to a reporter's question, he agreed that, yes, people should calm down with a glass of wine.

The argument that formed the basis of De Bernardinis and Bertolaso's reassurances—that small earthquakes reduce the risk of big earthquakes—is patently false. It is a bit of folk wisdom, one that I am asked about frequently, and that arises from nothing so much as wishful thinking. Big earthquakes release more energy than small ones. If we have a lot of small ones, the argument goes, shouldn't that release the pent-up energy? While it makes intuitive sense, it contradicts the most consistent feature of earthquakes we've observed—one that Charlie Richter saw in the first set of earthquakes for which he calculated magnitudes, that we see in every aftershock sequence, and that occurs in any regional grouping of earthquakes around the world. *The relative number of small to large earthquakes is constant.* More small earthquakes means *more* big ones. Mathematicians call this a self-similar distribution.

With the Richter scale, this means that if there is one magnitude 3, we can expect approximately ten magnitude 2s. If there is one magnitude 6, there will be approximately ten magnitude 5s, one hundred magnitude 4s, and one thousand magnitude 3s. We may see small variations, of course. But this distribution is a truism in seismology. No seismologist would ever suggest that having many small earthquakes would make a bigger earthquake *less* likely.

So why did the Civil Protection authorities say that it did? We know from Bertolaso's taped conversation that he had this message in mind prior to the meeting. The Civil Protection authorities are the ones charged with the public protection. No scientists were at the press conference—it appears they were not invited. One said he didn't know it had happened until he returned to Rome. But why didn't the seismologists speak up afterward and say their spokesman was wrong?

*

The next week, late at night on April 5, Palm Sunday, a magnitude 3.9 struck the city. Dr. Vincenzo Vittorini, a forty-eight-year-old surgeon and native of L'Aquila, later spoke to *Nature* magazine about his reaction. "My father was afraid of earthquakes," he recalled, "so whenever the ground shook, even a little, he would gather us and take us out of the house. We would walk to a little piazza nearby, and the children—we were four brothers—and my mother would sleep in the car." When the magnitude 4.1 had hit L'Aquila a week earlier, his terrified wife had called him, and he had followed his father's lead, telling her to go outside and stay there for a while. But this time, he remembered the press conference, in which authorities had declared that a big one was less likely, an idea the town had been discussing in the intervening week. He, his wife, and their daughter debated what to do, and he eventually persuaded them to stay inside.

The three were together in the master bed several hours later, at 3:32 a.m. on April 6, when the big one hit, a magnitude 6.3 earthquake that tore L'Aquila asunder. While such a magnitude may seem small next to the earthquakes of the Pacific Rim, shaking directly atop a fault with an earthquake this size can be extreme. Essentially *every* building sustained damage—twenty thousand were destroyed altogether. The historic center, which had been rebuilt after the 1703 earthquake, was mostly destroyed. (It has remained off-limits in the years since, deemed too dangerous to enter.) Even more modern buildings, those erected in the boom after World War II, were mostly designed before the institution of seismic codes, and many suffered from substandard materials and construction. Dormitories at the University of L'Aquila collapsed, killing students. More than sixty thousand people were made homeless by the earthquake. The government set up refugee centers with tents that housed forty thousand of them. (The Italian prime minister Silvio Berlusconi did little to soothe sensibilities when he suggested that citizens should be grateful that the government was financing their stay, that they should treat it like a holiday at the beach.)

Dr. Vittorini described the earthquake as like being in a giant blender. His apartment building, constructed in 1962, collapsed altogether. His third-floor apartment came to rest just a few feet above the ground. He was pulled out alive six hours later. His wife and nine-year-old daughter died, two of the 309 victims of the destruction of their city.

*

As we have seen, every disaster provokes an impulse to blame. The suffering citizens of L'Aquila had an easy target in the government authorities and their unwarranted assurances. The government responded by convening an International Commission on Earthquake Forecasting for Civil Protection, just a few weeks after the disaster. They brought in ten leaders in seismology from nine countries, representing China, France, Italy, the United Kingdom, Germany, Greece, Russia, Japan, and the United States. The commission was chaired by Dr. Thomas Jordan, the former chair of earth sciences at MIT, then director of the Southern California Earthquake Center. He and his team performed a comprehensive review of earthquake prediction around the world, confirming that such prediction was impossible. Their conclusion, issued several months after the earthquake, was that the scientific community needed to take ownership not just of their research but of communicating effectively to the public.

While this went a way toward assigning "responsibility" for the destruction of L'Aquila, it proved insufficient. Seventeen months after the earthquake, the scientists and civil protection officers were blamed in a much more concrete way. In September 2011, the prosecutor for the state of Abruzzo issued an indictment charging Bernardo De Bernardinis, the vice director of Civil Protection, and the six seismologists and engineers who had participated in the fateful meeting on March 30 with involuntary manslaughter for their role in issuing false assurances to the public.

International science organizations responded with outrage. The American Association for the Advancement of Science, the International Union of Geology and Geophysics, and the Seismological Society of America, among others, sent letters to Italy condemning the indictments as an attack on science.

But L'Aquila never represented a failure of science. It was a failure to communicate the nuances of the data. In the face of tragedy, we always ask, "What could have been done differently?" In L'Aquila, there was a very obvious answer. "Distracted by Giuliani's predictions, the authorities did not emphasize this increase in hazard," Dr. Jordan of the international commission observed. "Neither did they focus on advising the people of L'Aquila about preparatory measures warranted by the seismic crisis. Instead, they were snookered into addressing a simple yes-or-no question: 'Will we be hit by a larger earthquake?'" By saying no, they presented a much more definitive, and in the end incorrect, statement.

The prosecution's case rested on the personal testimony of people like Dr. Vittorini. Another victim, Maurizio Cora, spoke of taking his family to an open plaza after the March 30 magnitude 4.1 event, but, because of the government's assurances, staying home after the magnitude 3.9 event on the night of April 5. His wife and both daughters perished in the collapse of his apartment building.

Their testimony was compelling, and the prosecution prevailed. All seven defendants were convicted and sentenced to six years in jail. Over the next three years, the convictions went through two appeals courts. The original judge had concluded that the experts had carried out a "superficial, approximate, and generic" risk analysis, and thus had failed to do their duties on the Risk Commission. An appeals court overturned this ruling based on the content of the commission members' analysis. It was a valid scientific opinion (even if not consensus across the field) to suggest, as their findings did, that there was no reason to believe the risk had changed significantly as a result of the earthquakes. In the appeal that followed, the prosecutor argued that the crime was in the scientists' failure to reject the "favorable discharge of energy" idea that Vice Director De Bernardinis offered up at his press conference. In the end, the court ruled that only De Bernardinis was at fault. His conviction was confirmed, with a reduced sentence of two years, while the scientists were acquitted.

*

While such a legal proceeding is highly unusual, the Italian scientists' predicament is not. My colleagues and I faced a similar one in California. Three weeks before the L'Aquila earthquake, a magnitude 4.8 earthquake happened just three miles from the southern end of the San Andreas Fault. The location is meaningful because earthquakes are more likely to trigger other earthquakes in nearby areas. If, as in this case, a very long fault is very nearby, this greatly increases the chance of triggering a *very* big earthquake. Almost two decades earlier, I had worked with a colleague from UC San Diego, Duncan Agnew, to create a methodology for estimating this increased hazard. We projected that a triggering earthquake such as the one we experienced in March 2009 would result in a 1–5 percent chance of a big (at least magnitude 7) earthquake on the San Andreas Fault occurring within the next three days.

The California equivalent of the Italian Risk Commission was the California Earthquake Prediction Evaluation Council, or CEPEC. I was serving on CEPEC at the time, and within an hour of the magnitude 4.8, we had convened a conference call. It took a couple of hours to come to agreement, so it was a few hours after the earthquake that our one-page statement of concern was given to the state of California. In it, we pointed out that the risk was low in absolute terms, but still hundreds of times higher than the long-term risk. We prepared a suggested public release with specific actions that could be taken by Southern California residents, such as checking their water supplies. All that was left was for the state to release our statement.

This process was the result of an agreement between the scientists of California and the Governor's Office of Emergency Services (Cal OES) that dated back to the late 1980s. Because Cal OES has to deal with the consequences of these statements, they understandably want to see them first and be their conduit to the public. So a protocol was worked out between the seismic networks, CEPEC, and Cal OES wherein the networks processed the data to determine what was going on, CEPEC crafted an assessment of the risk, and Cal OES released it to the press and the public. To keep scientists from going out on their own, Cal OES committed to sharing such information within thirty minutes of its transmittal.

This agreement was hammered out at a time when a lot of earthquakes were happening in California, from 1986 to 1994. We had plenty of opportunities then to see what worked and what didn't. But by 2009 we had entered a relatively quiet time, with few significant earthquakes. Governors had been voted in and out of office; officials, scientists, and technicians retired; relationships were lost. When we at CEPEC sent our statement to the Governor's Office that March, it went to a group of people who were by and large coming at such a report for the first time.

Nothing happened. No public statement was made. After a few hours, the scientists at the networks and CEPEC began to get antsy. A few more hours passed before word finally came back: Cal OES had decided not to make a public statement.

We argued about what to do. The risk of a subsequent earthquake dies off rapidly with time, just like aftershocks, so by the time we knew that Cal OES was withholding our statement, at least half of the added risk was gone. The U.S. Geological Survey, as a federal agency, could have made its own public statement, but we didn't have a mechanism in place to approve it—we had always gone through Cal OES. And so, in the end, no public statement was issued, and as we all know, no earthquake along the San Andreas Fault was triggered.

When just weeks later L'Aquila hit and scientists were held accountable, Thomas Jordan (who was involved in both, as chairman of the International Commission on Earthquake Forecasting and as a member of CEPEC) and I realized that we had dodged a bullet. If a San Andreas earthquake had been triggered, and if it became known, as it surely would have, that scientists had said there was a 5 percent chance of it occurring (because our 1–5 percent range would undoubtedly have been truncated) and that we had *not* informed the public, we would have been eviscerated. And we would have deserved it.

*

Science works only when its practitioners are free to argue opposing sides. Many of the Italian scientists caught up in the L'Aquila prosecutions will likely never again offer an opinion on the geologic hazard of a situation, and you can understand their hesitation.

But deciding to withhold such information for fear that it might be misunderstood—or worse, could be manipulated by the public—does the public a disservice. Disasters leave in their wake a void of information. If scientists don't fill it, someone else will. L'Aquila amply demonstrates the danger that such a situation presents.

Scientists labor in an unusual setting—one that assumes conflict and debate, where a certain amount of training and expertise is assumed. When we find that those outside our field fail to appreciate what we are trying to say, it is way too easy for us to assign *them* the blame, and not our failure to effectively communicate.

Scientific research largely takes place in universities and government laboratories. These settings reward breakthroughs—placing the best papers in the most distinguished journals. Time spent translating science for public consumption is time spent away from the work that's most likely to advance one's career. Meanwhile, there are the local governments, urban planners, engineers, and utility managers on the other end of the spectrum charged with acting on the latest science—to build safer cities, manage ecosystems, and protect our lifelines and transportation systems. They aren't funded to translate our research into application. The scientific endeavor has created too large a gap between the scientists and the people their work can best serve.

At the most fundamental level, we still need seismologists to research how earthquakes operate. Even today we don't know what makes one earthquake slip a few yards along a fault and stop, becoming a magnitude 1, and another slip for a hundred miles to a magnitude 7. We don't know if it's a product of something in the earth, before an earthquake begins, or if it happens dynamically, dependent on what the earthquake runs into (or doesn't) as it moves down a fault. If it's the latter, we'll likely never be able to offer the kind of prediction people crave.

For now, we have the one nonrandom piece of the earthquake puzzle—what we know about foreshocks, aftershocks, and triggering. We need to share such information with the public as clearly as we can. We need to trust the public with this information. The fear of misunderstanding leaves in the dark too many people who could use our insights to help themselves. Science can function only when its results are shared.

For now, I can say that there *will* be an earthquake in California today, and every day. How big it is will be anyone's guess.

CHAPTER ELEVEN

THE ISLAND OF ILL FORTUNE

Tohoku, Japan, 2011

If paradise now arises in hell, it's because in the
suspension of the usual order and the failure of most
systems, we are free to live and act another way.

—Rebecca Solnit, *A Paradise Built in Hell*

Great catastrophes are never caused by one factor alone. We can
cope with and respond to an isolated event. You might think of
it like the elements involved in a car accident: If one driver hadn't
been distracted by his children, and if another hadn't changed lanes
at that precise moment, and if a third car hadn't lost traction due to
the rain, there might be no incident to speak of. Remove any one
of the elements, and the moment can be withstood, corrected for.

We've seen the principle play out in societies facing natural
disasters. The earthquakes in Lisbon and Tokyo became catastrophes not just due to their massive shaking, but because of the
fires they triggered, compounded by an unfortunate time of day or
day of the year. What if the Lisbon earthquake hadn't struck during church services for All Saints' Day, if Tokyo hadn't hit at lunch
hour? New Orleans wasn't destroyed by Hurricane Katrina singly,
but by the subsequent failure of man-made levees. What if those
levees had been inspected by the U.S. Army rather than by a local
commission?

It's a familiar refrain now that natural disasters expose and apply pressure to weak points. They can result in profound systemic change. A forest in a climate that is getting hotter and drier can survive the stress for a while, until a wildfire burns it down so thoroughly that it simply cannot grow back. That ecosystem is gone, to be replaced by plants and animals better adapted to the new climate. The same holds for social systems, as these same disasters have demonstrated.

But alongside the devastation that an extreme natural event can bring, there are also opportunities. Disasters have contributed to the collapse of civilizations, but they've also been the catalyst for needed social change.

The Great East Japan earthquake of March 11, 2011, was just such an event—both an accumulation of physical and man-made forces that together led to catastrophe, and one that offered an opportunity—whose impact was so severe that long-standing tenets of the culture were, and still are, being reshaped. The women described in this chapter are emblematic of these changes; they contributed to their communities in ways they never could have imagined before disaster struck. The 2011 earthquake blew apart many of the constraints of their traditional culture and created opportunities for them, and others like them, to lead.

*

Because Japan is, at its core, a chain of volcanoes, flat space is a prized resource, forming only where rivers have, over millions of years, smoothed out ridges and valleys. On the main Japanese island of Honshu, a long valley runs north of Tokyo, creating a conduit for trade and a chain of cities that were historically the strongholds of various samurai clans. The city of Fukushima lies along this valley, two hundred miles north of Tokyo, close to the heart of Japan. Its name is translated as Island of Good Fortune, and indeed it was: until March 11, 2011, Fukushima was a thriving, prosperous community.

On that day, Maki Sahara was in her home in Fukushima City. Slender and tall with long bangs and shoulder-length hair, this young housewife was looking forward to her daughter's graduation from preschool the next day. In Japan, the school year ends in March, and preschools have a formal graduation to mark the

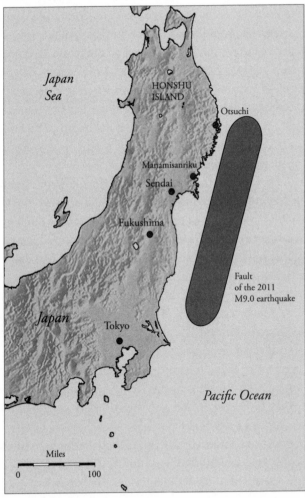

Map of Japan showing the fault for the 2011 Tohoku earthquake

children's entrance to a new stage of life. Two days earlier, a magnitude 7.2 earthquake had rattled the home, but an earthquake of that size strikes once or twice a year somewhere in Japan, and so, accustomed to the disruption, Maki didn't pay it too much mind.

Maki was going to be the representative of the Parents and Teachers Association at the ceremony, and she recalls laying out the kimono she would wear. A traditional kimono is a complicated ensemble, and she needed to get all the pieces in order. Maki remembers thinking not only of her six-year-old daughter at preschool, but also of her husband, who was at work at a nearby hotel. Her two nieces, staying at their grandparents', were also on her mind. Their mother was in the hospital battling leukemia, and Maki helped look after them while their mother was in treatment.

At 2:46 p.m., as the kimono lay on Maki's bed, the shaking began. A massive earthquake threw her to the floor. She braced herself, waiting for the shaking to stop. And waiting. Shaking so strong it was impossible to stand persisted *for over a minute*.

The earthquake was magnitude 9 and had occurred just offshore. The fault was about 250 miles long, centered due east of Fukushima. The earthquake was the fourth strongest ever recorded and had the largest slip ever seen. Until then, the world's largest slip had been the 1960 Chilean earthquake. That fault was 800 miles long, and it showed a maximum slip of about 120 feet. That means two objects that were adjacent on opposite sides of the fault would, in an instant, have been more than 100 feet apart. (Compare this to the mere 26 feet of buildup along the San Andreas Fault today.) The Japanese earthquake, though on a fault only one-third the length of the 1960 Chilean earthquake, still saw a maximum slip of 240 feet, twice the largest ever seen before. It was an earthquake most seismologists would have said couldn't happen—until it did.

It is a testament to the building codes that Japan had developed (and, just as important, rigorously enforced) that this earthquake, which would have leveled buildings in many other countries, created only a mess of broken dishes for Maki. Although her house was fine, the electricity was out and cell phone systems were too overloaded to use. First she ran to the preschool and brought her daughter home. Her husband came soon after, making sure his family was safe before returning to the hotel to take care of terrified guests. Maki and her daughter settled down, waiting for life to return to normal. But for Maki, life—as a mother, as a homemaker—would never be quite the same.

*

North of Fukushima, farther along the valley, lies the city of Sendai. While Maki was preparing for her daughter's preschool graduation, a thirty-five-year-old Canadian researcher in political science, Jackie Steele, was participating in another Japanese tradition to mark the six-month birthday of her infant daughter, Sena. She and her partner had brought Sena to a photography studio in a big shopping mall for a series of portraits in kimono, dress clothes, and even a bumblebee costume. They had just gotten Sena back into her own clothes and were starting to pick out which pictures they wanted to keep when the trembling began. They crouched on the floor, and, as the shaking continued and grew in intensity, the mall's power went out and darkness enveloped them. Jackie held her baby under her through the interminable moment, marveling at the calm of the employees around her. When the shaking finally stopped, the emergency lights produced a dull red glimmer and store employees led the customers, in the oft-drilled evacuation plan, to the roof of the building.

The roof was a parking area and Jackie's car was nearby, but she didn't leave. Once again the building sustained no significant damage, but now the people around her were in shock. Jackie recalls a man attempting to drive away who couldn't seem to remember what to do. He alternately raced the accelerator and slammed the brakes. His car seemed to stand up on its front. She didn't want to be on the road with people in this state, so she decided to wait it out on the roof, wrapped in blankets supplied by the employees of the mall.

Two hundred miles to the south, at the other end of the long valley, sits Tokyo, where thirty-eight million people fill the area that surrounds the Tokyo Bay. Megumi Ishimoto, a slight, energetic forty-year-old, worked in one of the city's many high-rises as an executive assistant to the CEO of a financial services company. Dissatisfied with her career, eager to do more than make money for investors, she was considering going abroad to engage in humanitarian work.

When the earthquake began to shake her high-rise, she wasn't thrown to the floor. Whereas Fukushima and Sendai both lay due west of the earthquake fault, close to the source of the waves, Tokyo lay south of the southern end of the earthquake rupture zone. The earthquake's waves had to travel farther to reach the city, which meant that while the shaking was very strong and lasted for almost two minutes, the sharpest jolts—the fastest, high-frequency shaking—had been dampened along the way. On the thirty-eighth floor of a Tokyo high-rise, those slow waves felt to Megumi like she was in a large cruise ship being tossed about on rough seas.

In the immediate aftermath of the earthquake, power was lost, and the city's vaunted train and subway systems were shut down. Millions and millions of commuters had to get home on foot. Megumi was lucky that she lived relatively nearby, and it took her only two hours to walk home; many of her coworkers walked for six or eight hours to get home and find out if their families were okay. All things considered, however, Tokyo did not suffer much damage.

At this point, Japan seemed to have withstood an impossibly large earthquake. Had this been the extent of it, it would have been a nation bruised but still sound—the driver momentarily distracted by his children, but who manages to avoid a collision; the forest diminished by a drought, but that ultimately continues to thrive.

The catastrophe that followed was the result of more than just this one event, and it had unforeseeable consequences.

*

It was six years earlier that the Sumatran earthquake had struck. As was revealed to horrified onlookers, an offshore earthquake's greatest damage is often inflicted not by the shaking, but by the resulting shift in the seafloor. The East Japan earthquake moved a 250-mile-long block of rock as much as 240 feet, displacing a great deal of water and leading, inevitably, to a tsunami.

Waves struck the northeasternmost part of Honshu, called Tohoku. The Tohoku region is a rugged coastline punctuated by small towns, most of which are fishing towns that harvest the seafood that Japanese cuisine is known for. Rural and isolated, this area is one of the more traditional parts of Japan. Eldest sons inherit farms, remaining in the family home with their parents. Wives live with their in-laws, and young mothers are expected to keep small children out of public view, creating a very limited life for women.

Tsunamis, like earthquakes, are part of life in Japan, and most towns had established defenses. In March 2011, at the time of this event, many had built seawalls up to twenty feet high to protect the flatland around the harbors where these towns had sprung up. The tsunami would take fifteen to thirty minutes to reach the shore, and residents had been trained to go to high ground after such a strong earthquake. They were aware of the danger and thought they were ready.

And they *were* ready for the tsunami that seismologists expected. But this earthquake exceeded all predictions for what the offshore fault could produce. The actual tsunami was several times larger than expected, at many places over forty-five feet high. One location saw waves one hundred feet high. It simply overwhelmed them. Alongside the northern half of the rupture, where the largest slip on the fault occurred, wave heights were *all* above thirty feet. Tide gauges in that section were destroyed by extreme waves—we can't know how large they were, only that they were larger than the height at which the instruments broke.

In the Tohoku town of Otsuchi, Takuya Ueno lived with his parents in the home his family held for generations. A thirty-three-year old salaryman, Takuya was one of the few people in town with a university degree. Otsuchi was a small community of sixteen thousand people, many working as fishermen or in fish processing plants. As part of the management team for a manufacturing center, Takuya commuted each day to the largest city in the region, forty miles to the north. As soon as the shaking stopped, he and his coworkers headed uphill. They watched the tsunami sweep through the city below, surge after surge washing upon it for hours. He and his colleagues survived, but they were stranded. The roads Takuya needed to drive home were destroyed; he had no way of getting there. He worked his way to an uncle's nearby house and waited.

Takuya's mother, Hiro, had spent the day at a medical clinic back in Otsuchi, helping her diabetic brother get care. She didn't drive, so her husband had dropped them off at the hospital. After the earthquake, knowing the tsunami risk, the patients and staff were all taken to the roof of the building, a practice called vertical evacuation. (It would have been impossible to get that many elderly and infirm people to safety in any other way.) The building was just tall enough to exceed the waves, and the patients survived, watching first the waves and then the fires destroy their homes.

Others in Otsuchi weren't so lucky. The city council met in an emergency session at city hall immediately after the earthquake. Their protocol had required them to leave city hall and move to high ground, but they decided to stay in place. The magnitude 7.2 earthquake two days earlier had resulted in a tsunami warning, but nothing had happened. The warning for this earthquake said to expect sixteen-foot waves. City hall sat behind a twenty-foot-high seawall that seemed more than enough to ensure their safety.

Many citizens gathered outside to hear what was being decided. When a forty-five-foot tsunami came pouring over the seawall, they ran inside to try to get to the roof, but the only passage up was by a single ladder. A handful survived, while hundreds of others were swept to their deaths, including the mayor and most of the city council. Of the sixteen thousand residents of Otsuchi, thirteen hundred lost their lives. Hiro and her brother were taken to an evacuation center, where she could only wait for someone from her family to come for her. Their home was near the ocean, and the waves had swept through her home and taken everything away. It was completely gone.

Three days after the tsunami, the roads still closed, Takuya heard that some people were fleeing Otsuchi using an old mountain path. Following it in reverse, Takuya made it back to Otsuchi and found the evacuation center and his mother, who screamed and collapsed when she saw him. His father hadn't returned.

The community struggled to come back from the brink of destruction in the weeks that followed. Emergency responders from the national government set up evacuation centers. Debris from the tsunami was collected. Bodies were sent to government-funded morgues and sorted by gender, age, and size. Takuya's close friend had lost his mother, and so, each day, the two went together to all the morgues, Takuya looking at the male bodies and his friend at the female, each trying to find their parent's remains. Each day, along with hundreds of others, opening body bag after body bag, examining the dead, hoping against hope, searching for closure.

A month after the tsunami, Takuya's father was finally found and brought to the morgue, where Takuya identified him. He had been found in his car, perhaps trying to evacuate or heading to the hospital to find his wife. After a month, in a mound of tsunami debris, he could be identified only by his distinctive watch. Four hundred of the dead in Otsuchi were never found.

Similar devastation played out in town after town on the Tohoku coast. In Minamisanriku, a young woman stayed at her post on the third floor of the emergency services building broadcasting tsunami warnings, giving instructions. The tsunami ripped through the building and carried her to her death. In an elementary school in Ichinoseki, the teachers, untrained in how to respond to a tsunami warning, kept the children in their schoolyard. More than two miles from the ocean, they thought they were safe. The tsunami swept through despite the distance, killing 74 of the school's 102 students.

In the end, 150 deaths were caused by the magnitude 9 earthquake, and more than 18,000 people were killed by the tsunami. Each of these deaths is a tragedy, and if this were the extent of the damage, it would be a significant, terrible disaster. Still, for an earthquake and tsunami of this unprecedented scale, Japan had weathered it as well as a country might reasonably hope to. The earthquake collapsed few buildings, and no trains were derailed. The tsunami, so much larger than expected, killed in terrible numbers, but its physical effects were endured by a relatively small part of Japan's population of more than 100 million, many fewer than the 140,000 who had died in Tokyo in 1923. What turned this event into a national catastrophe was not the earthquake alone, or even the pairing of earthquake and tsunami, but these two natural events in conjunction with one other significant man-made factor. Together they created a crisis on a scale that Japan hadn't seen since World War II.

*

Nuclear power plants harness the huge amount of heat generated by the splitting of the nuclei of large atoms, like those that make up uranium. This heat is used to create steam, which in turn is used to drive electric turbines and create the energy many of us consume in our daily lives. But the heat created in these nuclear reactions needs to be managed, to be drawn away from the nuclear fuel, or else we risk it melting, which will cause the vessel that contains the fuel to explode. So nuclear fuel is kept in circulating water to dissipate its heat. As a consequence, nuclear power plants are always built near major water sources, often the ocean, to facilitate cooling. Additionally, they have redundant backup power systems to make sure the heat management systems don't fail.

In Japan, because of the ongoing risk of earthquakes and tsunamis, ocean-side installations must consider the probable maximum tsunami they're likely to face. When the Fukushima Daiichi plant was built in the 1960s, the "design basis tsunami height"—the maximum-height tsunami they might expect—was ten feet. The plant was built thirty-three feet above sea level, which seemed a very comfortable margin of safety, and the engines for seawater pumps—the pumps that bring in water to cool nuclear fuel—were put at thirteen feet above sea level. In 2002, the design height was revised to twenty feet, based on more detailed studies of past tsunamis, and the seawater pumps were sealed accordingly, to protect them from inundation. In the years that followed, ancient records were studied that revealed an earthquake that had struck in AD 869, and that may have led to an even larger tsunami. In 2011, that risk was still being discussed; the appropriate action had not yet been agreed to. The Earthquake Research Committee was planning to release a report on it in April.

The Fukushima Daiichi plant withstood the earthquake itself without significant damage. The nuclear reactors were automatically shut down, as they were designed to do. But even once active reaction is stopped, there are residual processes that generate heat, and cooling remains essential. The earthquake caused a failure of the electric power grid that ran the cooling pumps, so backup generators kicked in. They seemed to be working as they should.

The first surge of the tsunami hit Fukushima almost an hour after the earthquake, at 3:41 p.m., with a second, even larger surge eight minutes later. The seawater pumps, whose engines had been sealed a decade prior, could withstand the waves, even at the massive scale they were experiencing. It was in the backup generators that the vulnerability lay. They were at too low a level and were completely inundated by the forty-foot-plus waves. As a consequence, the cooling systems failed for three of the six reactors on the site.

Without effective cooling, the reactors overheated. Pressure built and nuclear fuel melted. It was only a matter of time before reactors began to explode.

The earthquake and resulting tsunami occurred on a Friday afternoon. That evening, the Japanese government issued an emergency declaration, evacuating people living within two miles of the plant. On Saturday, the evacuation zone was extended to six miles. Then an explosion caused by the melting fuel in Unit 1 blew the roof off the reactor building, releasing more radiation, and the evacuation zone was extended to twelve miles. On Sunday, a water injection system failed in another reactor, and the water level began falling dramatically as well. Although it wasn't clear at the time, damage to the nuclear core had begun early that morning, and much if not all of the fuel in at least one unit had melted. On Tuesday, another explosion led to an expansion of the evacuation zone to twenty miles.

The plant had largely withstood the earthquake and even the tsunami, but those emergency generators were its Achilles' heel. The insufficient cooling led to nuclear meltdowns, hydrogen-air chemical explosions, and the release of radioactive material from three different reactors over the next four days. These incidents caused radioactive material to be released into the air and surrounding ocean. The chemicals emitted *ionizing radiation*—radiation with enough energy to change the atoms it comes into contact with. When that radiation hits human beings, it can alter our cells, leading to birth defects, cancers, and, in the highest doses, radiation poisoning and death.

*

Maki Sahara's home in the city of Fukushima was a bit more than thirty miles from the Fukushima Daiichi power plant. She and many of the almost three hundred thousand other residents of the city knew nothing about the problems at the power plant until four days after the tsunami. Without power, she had no television, and she and her neighbors were still busy dealing with the earthquake and its aftershocks. Her daughter's preschool graduation had been postponed as they waited for electricity to be restored. Residents were cleaning up the mess left behind by the earthquake. Food thrown out of cupboards onto the floor, broken dishes, glassware and knickknacks—it all had to be dealt with. It was on Tuesday, when evacuees from the coast started to arrive in Fukushima, that Maki and her family discovered that this was more than just another large earthquake. These people, who had been forced out of their homes, some arriving with just the clothes on their backs, were measured for radiation exposure. Some of them carried such substantial radiation that even the clothes on their backs were taken from them.

The government had first described the severity of the nuclear incident as a "4" on the international 0–7 scale. For reference, the Three Mile Island incident in the United States in 1979 had been a 5; the Chernobyl incident in the Soviet Union in 1986 was the only 7 to have ever occurred. The government warned that there was the possibility of more reactor explosions, but that the radioactivity released into the environment did not pose a risk to human health at that time. Maki noted the phrase "at that time" and wondered what was coming next.

Over the next month, as the nuclear crisis unfolded, they discovered that the situation was much worse than they'd been told, especially for residents of Fukushima. The reactors and spent fuels continued to overheat, requiring them to be bathed in untreated seawater. Seawater is corrosive, and leaks developed. Fires broke out, carrying more radiation into the environment. Storms and wind currents carried the radiation to the northwest—toward Fukushima. Radiation levels in the vicinity grew. After two weeks, even the tap water in Tokyo, over 150 miles south, showed twice the safe level for infants. By the end of March, radiation levels in the seawater immediately adjacent to the plant had levels of radiation that were 4,385 times what was considered safe. It was only with this absolute proof of the severity, a full month after the incident began, that the Japanese government revised the rating of the incident to a 7—on par with Chernobyl.

As the circle of danger widened, many residents of smaller towns near the power plant moved to evacuation centers in Fukushima City. The city itself was a hot spot—it showed the highest levels of radiation outside the defined evacuation zone. But with three hundred thousand residents and now a wave of refugees, its evacuation would have been a herculean task. The residents were told it was safe to stay. Elementary school students were not allowed to play outside, but as the city regrouped, middle schoolers and high schoolers continued with their outdoor athletics.

But the radioactive materials were continuing to be released and dispersing into the environment. The heavier particles, like cesium and iodine, responsible for much of the radiation, settled into grassy areas and sandboxes. By the end of summer, the radiation levels in Fukushima City were so high that the government decided to remove the topsoil from every piece of exposed ground in the city. Schoolyards, parks, and backyards all had the top few inches of soil scraped up and put into large, sealed plastic mounds. It took five years to complete the removal.

The initial government assurances, followed by incidents that revealed those reassurances to be misplaced, sowed mistrust among the citizens. Maki Sahara was outraged at the lack of information and traveled to Tokyo to join protests. In those first few months after the tsunami, as many as two hundred thousand people took to the city's streets in demonstrations protesting the continued reliance on nuclear energy and the lack of communication about the situation in Fukushima. The antinuclear movement has a long history in Japan, the only country to have ever lost a city to an atomic bomb. The Fukushima Daiichi failure brought new focus to the movement, and activists came to Fukushima to help in the recovery. The Fukushima 30-Year Project was initiated, to help residents get information about the radiation. With donations, the project bought scanners to assess food for safety, as well as full-body scanners to measure absorbed radiation. Classes and training were provided so residents could learn how to protect themselves and their families. The project pressured the government to install public radiation monitors in parks, which finally happened two years after the earthquake.

Maki volunteered to help. She began as a receptionist, helping citizens access the services they needed. She didn't have training, but she wanted to make a difference. She scheduled the scans, signed people up for classes. If a scan showed a person had high levels, he would be offered counseling on ways to minimize exposure. The project offered classes that, for example, instructed parents to have their children play on concrete, not on grass or in sand. As time went on, Maki took on a larger and larger role. A local mother trying to protect her child was a much more effective voice for change than those of activists from Tokyo. She also knew her community and what the people needed. For instance, she was able to get handheld Geiger counters and created a training class to show children how to measure radiation, teaching them to recognize safe places to play. She empowered children to understand and take control, much as she had done. She found her purpose.

*

Takuya Ueno, like most everyone else in Otsuchi, had lost almost everything of his former life. His childhood home, the home of generations of his family before him, was completely gone, swept away from its place near the ocean. The government decreed that the site was at too much risk from future tsunamis for rebuilding. His job had vanished; the factory had been destroyed. His father was gone too, and his mother consumed by grief.

The national government came in with aid, but there was no local government to work with them. It took until August before the town was organized enough to elect a new mayor and city council to replace those who had died in the tsunami. How do you restart when nothing is left? Takuya began meeting with others who wanted to aid in the reconstruction. They brainstormed ways to create job opportunities.

Outside rescue workers came to help. Among them was an emergency nurse, Mio Kamitani, who had worked at a hospital in Galveston, Texas, in 2008, as Hurricane Ike put most of that city underwater. There, she had attended to patients too ill to be moved in the evacuation. Mio came to Otsuchi to provide psychological care, and she stayed to help the community rebuild. Takuya and Mio fell in love and married, living in the temporary housing units that were the town's only homes.

One of the most difficult tasks facing Otsuchi was grappling with the emotional impact of the disaster. Four hundred of the lives lost there were lost completely, their bodies never recovered. The trauma of the event, coupled with the sheer magnitude of the death toll, made the transition through grief particularly difficult. When *everyone* is suffering from PTSD, who is left to help? Several years after the tsunami, the town organized a trip to the sacred mountain of Osorezan. A desolate, volcanic landscape of sulfur-stained rock, the Osorezan site has, for more than a millennium, been venerated in the Buddhist tradition as a gateway to the afterlife. Many of the families who had lost husbands and wives, children and parents, went on this trip to pray for their dead and try to find closure. Takuya's mother, Hiro, went to pray for her husband and to let go of the grief.

Otsuchi residents looking to find a way forward first self-organized, and their efforts led to the creation of a formal nonprofit called Oraga Otsuchi Yume Hiroba, which could be translated as "A Field of Dreams for Our Otsuchi." The goal of the organization is to inspire reconstruction; provide support to supplement government functions that have been degraded or lost; and revitalize local industries and tourism. They offer workshops to other communities on how organizations can work together to bring a community back from the brink, using their own hard-won examples.

Takuya and Mio still ponder how an event could be so horrible and yet could be the event that brought them together, that created for them a new life.

Mio also derived from it a newfound sense of perspective. Asked what message she would share with others, in the wake of the catastrophe she witnessed, she said, "Love, thank, and cherish your loved ones every day of your life. . . . It sounds like a cliché, but that is what many cannot do anymore and miss the most. . . . My town and other disaster areas are talked about in the sense of disaster prevention/preparedness and reconstructions very frequently, but I think we can also be the ones to talk about 'love' because many of us have found our own definitions of what that is."

She likewise urged others to "believe that you have power and ability to make your own decisions," citing the dependence of many Japanese in 2011 on governments and agencies—groups that fell short in preserving their citizens' safety. "I think it is very important for people to know that it is they who make decisions for their lives, not the warning systems."

*

Megumi Ishimoto in Tokyo felt the effects of the nuclear meltdown more indirectly. All nuclear power plants were shut down, which meant that the entire country was running out of electricity. For many days, she went to work without power, without lights, without heat. She knew others were suffering more than she was, and she wanted to do something to help. But at that point there were few organized volunteer efforts. She and her friends went to Ishinomeki, one of the towns heavily damaged by the tsunami, to remove debris.

It was a turning point for Megumi. She came back to Tokyo, quit her job, and started looking for opportunities to do the humanitarian work that had been calling out to her even before the disaster. In early May, she went back to Tohoku, to another seaside town called Minamisanriku. From across Japan, tens of thousands like her were going to Tohoku to volunteer their efforts, and several different groups had formed centers to coordinate with local and national governments. Most volunteers wanted to do the physical work of clearing debris because it let them connect with the local people.

For her part, when Megumi arrived at the center in Minamis-anriku, she offered to do whatever was needed. With her executive management background, they asked her to take on the less glamorous back-office work, organizing teams and managing support for volunteers. She had originally intended to stay for a week, but the others begged her to stay longer. She agreed to stay for one to three months. The local government had requested that volunteers create a support group for the women in the evacuation center, and with her longer time commitment, Megumi was tasked with heading it up.

Together with a city government official and a local woman, she visited evacuation centers. At first, much of their work was simply listening to what these women most needed. Many of the women they interviewed had been raised to remain quiet, to focus on their families' concerns and not their own. In situations where the conditions in the evacuation center were rough, exacerbating the trauma of being forced from their homes, these women didn't have a cultural framework that encouraged them to protest. And so Megumi's first job was to provide an environment where these traditional women could feel free to speak up. She created knitting groups, offering them a reason to show up. The first few were very quiet. After some time, women began to speak up. They told of the difficulties of being in such close quarters, and of the older men angry that the women's young children made so much noise. They talked about the male manager of supplies at the evacuation center who handed out menstrual pads one at a time, and their embarrassment at having to discuss such a private matter with a strange man several times a day. And a few, very tentatively, began to talk of the sexual aggression they faced in the centers.

Through these conversations, Megumi came to realize that if her organization was going to help the disaster's most vulnerable—the babies and the elderly—they needed to ensure that the women were cared for, since they were the predominant caregivers. She also realized that as families were moved from evacuation centers to temporary housing, the city program, based in evacuation centers, was losing contact with them at a time when they still needed support.

Recognizing the gap, she decided to form a separate, farther-reaching women's resource center. It began with no money, but Megumi was able to secure funding from foundations, and gradually from government programs as well. It began as a place of support for women in temporary housing and otherwise affected by the tsunami—to talk, to assist in the struggle with bureaucracy. Her organization has evolved over five years into Women's Eye, a nonprofit dedicated to supporting the women in Tohoku as they work to rebuild their homes and communities.

The members are women entrepreneurs, creating businesses and nonprofits to revitalize Tohoku. Maki Sahara and Mio Kamitani are both members, as are a midwife starting a chain of modern birth clinics, a photographer, and the owner of a seaweed processing plant. Women's Eye connects them with one another to help them see they aren't alone, and it provides business and leadership training.

It also allied itself with a larger national movement, the Japan Women's Network for Disaster Risk Reduction. Akiko Domoto was the first woman governor of Chiba Province. As president of the organization, she tackles real problems in disaster response as it affects women, but also the underlying issues of gender inequity that disaster exposes.

For Jackie Steele, the Canadian researcher in Sendai with a six-month-old baby, the earthquake meant having to leave her home. Without heat and water, she didn't know how else to keep her baby safe. Unlike Maki, she heard about the nuclear problems as they unfolded, and she knew that her home was downwind from the plant and babies were the most at risk. Her parents and friends begged her to come back to Canada, but that would feel like abandoning her community in Sendai, and the two years of postdoctoral research in political science that were nearly complete. Still, after two very cold nights, she had no choice but to go *somewhere*. She had a half tank of gas, luckily, so she was able to drive away. She stayed with friends in Nagano, safely removed from the crisis.

Jackie ultimately left Sendai, but not Japan. Before March 11, Jackie's research had been on diversity and women's political citizenship. With her experience of the earthquake and her observations of response and recovery, she became interested in the governance of natural disasters—how governments function in times of crisis. She was especially concerned with how women were treated in this process, which is how she became connected to Governor Domoto's organization, as well as to Megumi and Women's Eye.

Jackie is now an associate professor of political science at the University of Tokyo, and she comes back to the Tohoku region to study how residents are creating something new in Japan. They are learning that these women are more than "just" mothers doing what they must to support their families. They are critical pieces of their communities, bringing it back to life, and creating a new, more inclusive future for women.

*

In the spring of 2017, I spent a day with Maki Sahara, learning about her efforts to bring radiation data, awareness, and training to Fukushima. She had by then taken over operation of the Fukushima 30-Year Project—a long way removed from her life as a housewife six years prior. She has the determination to keep the project going, to maintain focus. She knows one of the most difficult parts of recovery is the failure of our attention spans. The world inevitably moves on to other disasters, other crises, other needs. But for the people of Tohoku, the recovery was ongoing. Years later, many still lived in temporary housing. The area immediately around the

Daiichi plant was still uninhabitable. The town of Otsuchi was still trying to decide whether their devastated city hall should be kept as a memorial or torn down so the community could move on. Recovery can be an agonizingly long process.

At the end of our day together, I asked Maki if there was one thing she could share with the world, what it would be. She told me she wanted to be able, in twenty years' time, to look back and be relieved that she and her fellow organizers did *more* than what was required to keep children safe. Because it would be too awful to contemplate looking back and seeing that they had done too little.

CHAPTER TWELVE

RESILIENCE BY DESIGN

Los Angeles, California, sometime in the future

With each passing year the enviable physical, economic,
and social circumstances of the United States are more
vulnerable to natural and technological hazards. . . .
[The United States] has been—and still is—creating for
itself increasingly catastrophic future disasters.

—Dennis Mileti, *Disasters by Design*

Los Angeles owes its existence to earthquakes. Its location, in the
arid southwest, could have left it an uninhabitable desert had it
not been for the mountains that surround it, pushed up by active
faults, capturing moisture from the clouds that come off the ocean.
Those same faults trap groundwater, creating the springs that were
used by the original settlers to irrigate their crops. The modern city
began to flourish at the beginning of the twentieth century with the
discovery of oil—oil that was likewise collected by faults, with the
biggest deposits near the Newport-Inglewood Fault, which runs
through Long Beach and Los Angeles's Westside.

Faults may have made Los Angeles a viable city, but they are a
precarious asset, and the risk of earthquakes is ever present. The
Newport-Inglewood Fault was responsible for a magnitude 6.3
earthquake in 1933, destroying more than seven hundred schools

and shaking out with it America's first piece of earthquake safety legislation. The Field Act mandated higher building standards for public schools, the first seismic building codes. When the 1971 San Fernando earthquake destroyed not just decades-old structures but modern, engineered buildings—most significantly the new psychiatric ward at the Olive View Hospital—it led to a wholesale revision in the building code. When the 1994 Northridge earthquake struck, bringing down two freeway bridges, it convinced us that our freeways couldn't be trusted and triggered $10 billion in retrofits by the California Department of Transportation.

These three earthquakes, 1933, 1971, and 1994, have evoked in us contradictory beliefs. On one hand, they inspired us to ready ourselves—to ensure that we're suitably prepared to withstand and respond. But just as important, they have fooled us into thinking that we know how to handle earthquakes. After all, the city responded and thrived each time. In 1933, with seven hundred schools destroyed, students went to school in tents for two years, but all the schools were eventually rebuilt in a new, stronger style. In 1971, 110 fires were triggered and successfully fought. A dam almost failed, which would have inundated eighty thousand residents—but everyone was evacuated and the water level was drawn down in time so that no flood ensued. In 1994, the whole city lost electric power—and got it back in twenty-four hours. Each event triggered further safety legislation, with ever-stricter codes for schools, hospitals, and freeways. Each made us think that earthquakes were manageable, that the risk to Los Angeles was not so great.

But there are two dangerous flaws to this way of thinking. The first is that the resulting programs have always been designed to satisfy the public's outrage, keeping in mind the desire not to spend *too* much of the city's money. For each piece of legislation enacted, there are countless others that were proposed and that were just as worthy, but that couldn't muster the support to justify their price tag. The easiest programs to initiate tended to support response rather than prevention. The public can embrace the idea of a fire department that is sufficiently outfitted. But demanding that building owners repair their unreliable structures is a more difficult proposition. After all, they have to live with the consequences of their decisions. Shouldn't they be entitled to make their own choices, no matter how reckless?

This points us to the second major flaw in this perception of safety—namely, of course, that these earthquakes (1933, 1971, 1994) just *weren't that big*. They certainly seemed like it when you consider the cost ($50 million, $500 million, $40 billion) and the lives lost (115, 64, 57). But their magnitudes (6.3, 6.6, 6.7) reflect the relatively short faults that broke in each, none more than a dozen miles in length. Nowhere near the San Andreas quake my colleagues and I modeled. They weren't the Big One.

When Big Ones strike, our individual choices won't exist in isolation. The fires that swept through Tokyo weren't confined to the properties where they started. The water pouring out of a levee crevasse doesn't recognize county borders. Big Ones have the power not just to impact communities but to utterly transform them. They can destroy industries, as the California floods of the 1860s did. They can create a nation of refugees, like Iceland after Laki. They can set back economies for decades or more, as with eighteenth-century Portugal. They can make or break political fortunes. Preparing for the Big One is categorically different from preparing for the sort of big ones.

*

Eric Garcetti, mayor of Los Angeles since 2013, is the quintessential Angeleno. In a city of immigrants, he is his own melting pot. His Italian forebears settled in Mexico. His paternal grandfather was brought to California as an infant after his father was murdered in the Mexican Revolution. His mother was a Russian emigré, making him both the first Jewish and the second Latino mayor of Los Angeles. He often relates that he was born in the San Fernando Valley, near the fault that produced the 1971 San Fernando earthquake, just five days before it struck. He tells that story to suggest that he was perhaps destined to deal with earthquakes.

At the urging of a friend and colleague who knew Garcetti I met the mayor a few months after he took office. I wasn't sure it would be worth the time, frankly. I had led the creation of the ShakeOut scenario six years earlier, thinking that if we put our scientific understanding into concrete terms, people would recognize that outcomes were under their control, and thus feel compelled to act on them. The scenario was popular, widely read, and used by many agencies to plan their response to a big earthquake. But it wasn't being used for prevention the way I thought it should. People didn't seem to be able to make the leap from understanding the potential for damage to believing their actions could actually prevent it.

There was a ray of hope, though: a program slowly getting under way in San Francisco. A ten-year push by engineers and scientists in the area had finally led to the creation of the Community Action Plan for Seismic Safety, a blueprint for measures that could be taken by the city to reduce its risk in the event of an earthquake. So, with their initiative in mind, I requested a meeting with Mayor Garcetti to tell him about the San Francisco project and maybe get a little city rivalry going.

The mayor later told me that my visit was both exhilarating and terrifying. He was still new at the job. He told me that he had imagined that moving from his six years as city council president to mayor would be like trading up from a Toyota Corolla to a Camry—a similar job with better tools. The reality, he said, was more like going from a Corolla to a semitrailer truck. He was still trying to comprehend all his responsibilities. Then I arrived, attempting to convey with certainty and clarity what lay ahead, albeit at an unknown time in our future. It was, understandably, a lot to digest. It was as though I pulled him out of the truck and plopped him down in front of the controls of an airplane. The most fundamental function of government is to provide for public safety. During that meeting, I told him that guaranteeing public safety might be unattainable.

But he didn't shut down the meeting. We kept on talking. I found in Mayor Garcetti an elected official who thought more like a scientist than I had expected. He was committed to data, to evaluating what was being done in quantifiable terms. One of his big initiatives as mayor has been to open up the city's data, recognizing that sharing it was the only reliable path to improvement—"embracing the shame, if need be," he added.

We found that we understood each other, two born-and-bred Angelenos who wanted the same thing—to see our city endure. So we tried something that hadn't been done before. We negotiated an agreement between the U.S. Geological Survey, my federal agency, and the City of Los Angeles, in which I would spend a year at city hall, working with the mayor's staff to come up with solutions to the city's most pressing seismic problems. We made a big, public statement that we were collaborating to find solutions we couldn't have arrived at on our own.

What followed was a dissolving of boundaries. For me, it was a yearlong experiment in understanding not just what data was most relevant to the city in the event of the Big One, but also what information best connected with citizens, inspired them to act. For policy makers, it was an experiment in diving into the technical. Engineers came to city hall; city hall visited the Water Department.

I received my first lesson in the political realities of disaster prevention early on—in fact before I even started. The agreement between the USGS and the city that provided for my transfer required framing the problems we set out to solve in the year ahead. It did not call for a comprehensive seismic safety plan the way I might have imagined, but instead for addressing three specific issues—strengthening older buildings that we knew to be vulnerable; safeguarding the city's water system; and reinforcing our telecommunication systems. These were, without question, big and important problems to address. But they were far from the only ones. It was then that I realized how important it was not just to advance the cause more broadly, but to do so in a way that allowed us to demonstrate concrete, measurable levels of success. Only by being seen as successful would we gain the political support to see the project through.

Our message needed to register on the right emotional level, too. ShakeOut was the synthesis of decades of research, but to inspire action it needed to be a *story,* something that translated the science of the San Andreas earthquake into a tangible reality. When our research was published in 2008, we had created a short movie as a sort of synopsis, as well as a written narrative version of it (more accessible to the public, we hoped, than a scientific paper). The story we composed began ten minutes before the earthquake and ended six months after. To accomplish the mayor's goals, I relied heavily on both versions.

We decided to reach out to the unwitting actors in our story, those who stood to lose so much. I convened and presided over 130 public meetings in ten months. Meetings with building officials and building owners. With structural engineers and civil engineers. With apartment owners and tenants. With urban planners and urban developers. In addition to conveying the story of the earthquake that we had put together, I listened for reactions, ideas, solutions. Many details of the final proposal came from citizens who attended these meetings and would be affected by our plan. In addition to contributing valuable ideas, they were given a stake in its success.

I learned, too, to avoid all discussion of probabilities. The question of *when* a disaster will strike is perhaps inevitable, but it triggers our fear, compelling us to hide from the problem. *When* is a probabilistic uncertainty. As a scientist, I know that uncertainties matter—to convince my colleagues of a result, I need to demonstrate that I have analyzed them and taken them into account. But policy makers need to focus on the *what,* not the *when.* Their policies can't affect when a disaster strikes, but they can absolutely change its impact. I emphasized that the Big One was likely enough, that it would occur soon enough, to be worth our preparing for.

I made a point to emphasize financial consequences, not the threat to lives. Again, we were trying to steer people away from fear, reminding them that they would have to pay for the earthquake at one time or another, before or after, so why not avoid the damage altogether? We also focused on how our vulnerability is shared—that someone's decision not to be ready increases the chances that others in the area will suffer.

Our message got through. The mayor released our plan, called Resilience by Design, at the end of 2014. The culmination of my year at city hall, it included eighteen recommendations issued to address our three identified priorities, written with twenty members of the mayor's staff. It didn't solve everything, but it was easily the biggest step ever taken in California toward seismic safety.

It was the mayor's plan—he was the final arbiter of what was included. And even as we worked in close collaboration, I came to realize how important it is to maintain a distinction between science and policy. If scientists start making policy, we invite politicians to start making science. By instead empowering politicians with the information to make informed decisions, we create more forceful advocates for the results of our collaboration.

Some aspects of the plan were completely up to the mayor to enact. The city instituted a wide-ranging approach to protecting the municipal water system. New projects would be reviewed for their seismic resilience. Engineering plans were undertaken to protect the aqueduct that brings water from the Sierra Nevada Mountains and crosses the San Andreas Fault in an old wooden tunnel, built in 1908. The city committed to using seismic-resistant pipes to deliver water to houses and businesses; five such pilot projects are now installed. The Water Department is working with the Fire Department to create redundant, resilient sources of emergency firefighting water. A plan for citywide, solar-powered Wi-Fi, as a fallback for when cell towers run out of their four-hour backup power, is moving forward.

Many of the proposals required action by the city council. Several ordinances were proposed to retrofit two types of unreliable building, create a loan program for the retrofits, and require future cell phone towers to be built to a more resilient standard. Negotiations went on for almost a year, but in October 2015, the ordinances were passed by a unanimous vote of the city council. Many of the organizations representing the building owners, which might have otherwise presented significant opposition, had instead been part of the development process; they stood with the mayor as he announced the result. Even though building owners would have to pay the full cost, we had convinced them they had more to lose by not retrofitting. Perhaps more significantly, they had come to understand how much they stood to lose if their neighbor didn't do so. The fact that everyone had to do their part, and that the city would be paying for the improvements to the water system as well as the retrofits of their own buildings, made the sacrifice a shared one. Almost twenty thousand buildings will be retrofitted over the next seven to twenty-five years.

When our Big One comes, lives will be saved for what we collectively accomplished, a fact that never fails to amaze me. Scientific researchers rarely see such concrete results of their work. To my surprise, the mayor saw it the same way. "It is, to date, one of the best experiences of my life as a policy maker," he told me. "We did something incredibly complex, not easy financially, and we did it really with no opposition at the end, and in government terms, we did it very quickly."

The Los Angeles Downtown News, a regional weekly paper, remarked in an editorial on the political challenges of a plan like ours. "If the Big One strikes in the next few years, then there will probably not have been enough time to effect the most significant changes Garcetti wants. If a massive temblor hits after he leaves office, and the city turns out to be well prepared, then he may not be around to take the credit. All of which indicates that Garcetti may be focusing on earthquake safety because it's, gasp, the right thing to do for the city."

They're right on all counts. But in the funny, unpredictable way of politics, Mayor Garcetti has reaped benefits from the program after all. All the news outlets praised the program. And while this was obviously not the only activity the mayor undertook in his first term of office, he did win reelection to a second term with 81 percent of the vote. Other elected officials have noticed. The Southern California Association of Governments is supporting further seismic programs among its 191 other constituent cities. In the two years since Los Angeles passed its mandatory retrofit ordinances, two more cities, Santa Monica and West Hollywood, have followed suit and passed similar laws. When the Puebla earthquake of 2017 brought down many similar buildings in Mexico City, the *Los Angeles Times* reminded its readers of the courageous actions of Mayor Garcetti and encouraged even more cities to join. More than three dozen in Southern California now have programs in the works.

*

These Southern Californian initiatives, alongside the move to hire chief resilience officers in many cities around the country (many of them supported by the Rockefeller Foundation's 100 Resilient Cities program) and the UN's initiative for global disaster risk reduction, all point toward a broader awareness of disaster. We've seen it across the world in the last decade. It is a movement in which our scientific understanding—especially our ability to comprehend time spans longer than that of human memory—is allowing us to overcome our ingrained biases.

As civilizations, we have, across our history, come to disasters from a place of fear—of the unknown, of the unpredictable. We sought to find patterns in them. We developed culturally appropriate explanations—quarreling gods, divine retribution, celestial balances needing to be preserved—that allowed us to ascribe meaning to an event.

As our philosophical and ethical systems became more sophisticated, we grappled with the logical inconsistencies of such a view—how could a loving god kill the most innocent among us? A volcano as the expression of a cuckolded god's temper tantrum no longer sufficed. We turned to science to explain the natural world, to put these events in context. Now we see natural hazards as the result of variations in physical systems.

Our improved understanding of physical systems has shown us that many of the impacts of natural disasters can be reduced or eliminated through better design of the human systems that interact with the physical. Better management of floodplains, buildings that can resist strong winds or earthquake shaking, and warning systems for hurricanes and tsunamis all help protect lives and increase the capacity of communities to recover from disasters. Our focus on response, however, still makes listening to urban planners and building officials more difficult than, for instance, supporting our firemen. But I see even this starting to change. Where the immediate news coverage of Hurricane Katrina in 2005 emphasized the supposed breakdown of social order, the coverage of Hurricane Harvey hitting Houston in 2017 quickly turned to Houston's lack of zoning laws and how that had increased the damage.

The biggest shift in recent years, though, has been to move beyond a parochial view of our world. For the first time, a disaster on one side of the earth is helping motivate people on the other. Telecommunication gives us the ability to directly experience the suffering of others, deepening our empathy with the victims. Our last challenge is to see the victims, no matter where they are, as ourselves. In his book *The Expanding Circle,* the philosopher Peter Singer describes the evolution of ethics in the human species as an expanding circle of whom we include in our definition of "us." From self to family, tribe, nation, and eventually all humanity, we are broadening our definition of who deserves fair treatment and consideration.

In the summer of 2017, three hurricanes—Harvey, Irma, and Maria—hit the United States. All were extreme events with severe consequences, but they were not equal. Hurricane Harvey was primarily a flooding event, dropping more rain in one storm than ever before seen in the United States, with more than sixty inches in two different locations, Nederland and Groves, Texas. Well over one hundred thousand homes were damaged or destroyed, most of them without flood insurance.

Two weeks later, Hurricane Irma approached Florida. It was so large that all of Florida was hit with heavy rain and close-to-hurricane-force winds. The need to make sure that everyone in Florida was ready for the severe storm created messages that obscured how much worse the winds would be at the eyewall. Those few places actually hit by the eye of the storm suffered far greater damage. In the end, although the losses were significant, Florida got lucky. The eye of the storm skirted the western coast and the very worst winds stayed out of the most densely populated areas. Floridians may not have felt lucky, especially the many who lost their homes. But the final track of the storm, coupled with Florida's advance planning, meant that their disaster did not become a catastrophe.

The same cannot be said for Puerto Rico one week later. There we saw a truly catastrophic Big One that could change the society. Although Hurricane Maria was not as big as Irma in maximum wind speed or size, the eyewall traversed the length of the island, and the winds experienced on Puerto Rico greatly exceeded those that hit most of Florida. Paired with the damage from Irma the week before, to say nothing of a struggling economy and aging infrastructure, Puerto Rico suffered a loss of the basics of modern society for much longer than many people would have thought possible in this modern era.

The human response to the extraordinary hurricane season of 2017 gives us cause for cautious optimism. Whereas the initial news coverage of Katrina blamed victims, the coverage of Harvey tended to focus on the ways the community united, and how the unregulated expansion of impermeable surfaces made such a disaster possible. Houston is a multiethnic city, like New Orleans. It is clear that looting also plagued its disaster response. But stories of lawlessness didn't dominate this time. It encourages me to hope that our circle of empathy is growing wider.

The cautionary note, however, is that where the flooding in Katrina overwhelmingly affected impoverished neighborhoods, Harvey was more of an equal-opportunity assault, flooding poor and rich neighborhoods alike. Empathy is easier when you can see yourself in the victims. The initial response to Hurricane Maria and the devastation of Puerto Rico also suggested that empathy comes more slowly when the victims are Americans who don't speak English.

Empathy is a meaningful start. One of our greatest challenges as individuals, though, is moving beyond empathy to action— overcoming the sense of powerlessness that natural disasters inspire. Taking action, taking control is the best antidote to fear. If you have read this far, you may be thinking about what you can do to make your home and community more resilient to the risks before you. Here are some worthwhile first steps:

Educate yourself. Every city or town is subject to natural disaster in some form. Find out what risks your community faces and try to be rational about which are the most significant. The randomness and uncertainty of earthquakes, for instance, can inspire great fear, but in your community, flooding might be the more salient threat. Look at how scientists quantify the hazard, but remember that they are quantifying what the earth will do, and not what it will do to *you*. You can start by looking at the resources provided by the National Oceanic and Atmospheric Administration, for meteorological hazards, and by the United States Geological Survey, for geologic hazards.

It's just as important to weigh the actual damage your community could face. Much of it is preventable. The Federal Emergency Management Agency has resources describing mitigation strategies (ways of preventing your losses) for a variety of hazards. Your local, county, or state emergency services agency probably has both hazards and mitigation information for your region as well. The actions they suggest may cost money up front, but your efforts will almost always save you more in the long run.

Don't assume government has you covered. When it comes to the strength and safety of your house, apartment building, or place of work, don't rely on the government—for three reasons. First, the government doesn't pass building codes in order to protect you economically. The philosophy is that you're free to make careless investments; you just aren't permitted to kill yourself or anyone else in the process. Second, your building code is only as good as the building code in place when the structure was built. If you have a beautiful Victorian, it was built before *any* building codes were instituted. Third, for codes to work, they have to be enforced. An understaffed building department is a department that can't protect you.

If you own a building, you owe it to yourself to determine the risks it is subject to, and whether it is prepared to withstand them. Talk to a foundation specialist or a structural engineer (for a larger building). Find out what it will cost you to make it stronger. If you rent, talk to your landlord and fellow tenants. It could cost as little as $500 for work that could make the difference between substantial or even total loss of a home in a disaster and just mostly minor damage.

Engage with local leaders. Many of the most important actions a community can take happen in local government, as we saw in Los Angeles. But elected officials can do only what their constituents press for. If you care about stronger building codes, preservation of floodplains, or investment in safe infrastructure, you need to let your representatives know.

As you do so, remember that preemptive action tends to work best when it tries not to stop an ongoing physical process but to accommodate it. Attempting to stop river flow or sedimentation will always fail eventually. There is no mechanism for thwarting earthquakes, and there never will be. But a city's decision to build infrastructure as strong as truly necessary (and not just as strong as standing law might require) can save lives. It works best when we ask, "What outcome is completely unacceptable and what do we need to do to prevent it?"

Work with your community. Remember what's really at stake. You will almost certainly live through the disaster. Even in Pompeii, 90 percent of the residents escaped. It is the community, society itself, that is at risk. We know that damage occurs where a system is already weak. A community whose people know and care about one another is the one that will pull through. A community divided, whose ideas of preparedness involve procuring guns or fortified bunkers, is at risk. It becomes a self-fulfilling prophecy: if you treat your neighbor as a potential enemy, you make him one, and in so doing contribute to your society's collapse.

It is in the months and years after a catastrophic event, when the dust from disaster has settled, that the measure of a community is taken, when its future is tested. Those who go on to thrive do so because of individuals who make sacrifices for the benefit of others. Jon Steingrimsson held his community together in spite of his own great suffering and loss. De Carvalho inspired the king and his subjects to start rebuilding Lisbon before grief and despair could overwhelm them. In Japan, Maki Sahara moved beyond her life as a housewife to help mothers of Fukushima cope with their fears of radiation; Mio Kamitani left her previous life behind to lead her adopted home of Otsuchi to the future its residents dream of. It is not just elected leaders who guide us toward recovery.

Remember that disasters are more than the moment at which they happen. To effectively manage disasters, we must, as communities and individuals, focus on three different times: we must adequately build and retrofit our structures before the event, to minimize damage; we must respond effectively during the event, to save lives; and we must come together as a community after the event, to recover. Recognize that all three periods are important. Expand your definition of preparedness beyond simply *preparing to respond.*

And do this with your neighbors and friends. A church or mosque that has planned before the disaster, strengthened its buildings, and organized its members is an institution that can be a nucleus of recovery for the larger community after the event.

Think for yourself. It was overreliance on, and overconfidence in, existing engineering solutions that led Otsuchi's city council to meet behind their seawall after the Great East Japan earthquake, leaving them in the path of a tsunami. They put their lives in the hands of unknown scientists, instead of taking responsibility for themselves. Others can give you information, and you could and should do what you can to understand it. But ultimately the action has to be yours.

*

Natural disasters are becoming more common. As we have seen, heat—in the oceans and the atmosphere—is the fundamental driver of extreme storms, and the current warming trend is expected to increase both the number and spatial distribution of hazards. Even more significant is the expansion of our cities and the increasing complexity of urban life. Urban dwellers are ever more dependent on sophisticated supply chains for food, water, sewage, and power, even as we become more dependent on cell phones and the Internet for all aspects of our lives. The number of people who are vulnerable is growing rapidly. Whereas only 14 percent of the world's population lived in urban centers at the beginning of the twentieth century, over half of all human beings on earth live in cities, or almost four billion people. Many of these cities are located by the ocean, in tornado-prone areas, near faults, or at the base of volcanoes.

We need to accept that the timing of a disaster's occurrence is unambiguously random—we may never be able to anticipate the *when* of our big ones.

Human beings search for meaning in everything we do. On one level, it's a matter of self-preservation—it has inspired us to find patterns, to anticipate and predict future threats. But on a deeper level it reveals a desire for good to come of our actions. We need to recognize that our search for meaning can lead us to ask *why* such a thing might happen to us, when that impulse might instead be channeled into the question "How can I work with my neighbors to prepare and recover?"

The future is largely unknowable. We can see patterns and assess likelihoods, but time travels in only one direction. We cannot know which of the earth's many cities will experience their Big One in our lifetimes. But we can say with confidence that it *will* happen somewhere.

And when it does, in our globally connected world, we will all participate. We will share in the distress of the victims, as information is fed over our phones and computers. We will face our impulse to blame them, to try to understand what they did to deserve their misfortune. We will seek out a reason that might spare us from suffering the same fate. We will, in other words, experience the fear that stems from randomness. But we can acknowledge these impulses in ourselves and those around us and choose to move beyond them. We can acknowledge our deep-seated, instinctual responses to disaster, but draw instead from our enormous capacities for empathy, our willingness to help. We can use what we now know to help those most hurt by the disaster, and to prevent damage in the ones to come. Natural disasters strike us down together, and it is together that we will get back on our feet.

ACKNOWLEDGMENTS

This book is the product of a lifetime of experiences in my many, varied communities, too many to try to name, so I will acknowledge the most obvious and hope the others will forgive me. First, my agent, Farley Chase, sought me out and convinced me to try my hand at writing, helping me find the stories within the science. Then at Doubleday, I want to acknowledge the extraordinary guidance of my editor, Yaniv Soha, who helped me break out of the scientific mind-set and discover my storytelling voice, and the encouragement of his assistant, Sarah Porter.

The seeds of the book began with two scholars. My Chinese professor at Brown University, Jimmy Wrenn, introduced me to the approach to natural hazards in the ancient Chinese classics, after I discovered seismology and realized what my life's work would be. Marilyn McCord Adams, the Regius Professor of Divinity at Oxford University, an Anglican priest and close friend of my mother, helped me understand the evolution of ideas about natural disasters in the Judeo-Christian tradition.

From this base, many friends and colleagues contributed to my understanding of natural disasters. My husband, Egill, and his family in Iceland shared their rich traditions with me, and Alexandra Witze shared her research into the history of Icelandic volcanoes. Mike Shulters helped me understand the interplay of science and society around flooding in California, while Bob Holmes taught me much about hydrology and the Mississippi River. Peter Molnar got me to China, and many friends in China shared their experiences with me at a time when this was not without risk. Kerry Sieh and John Galetzka helped me understand not just Indonesian tectonics but also how field geologists think and work. My understanding of the social side of disaster prevention was honed through my extraordinary experiences in the Los Angeles mayor's office and the many people who participated in that process, especially Eileen Decker, Peter Marx, and Matt Petersen.

The story would have been incomplete without my many friends and colleagues who have shared their personal experiences, especially Donyelle Davis, Andreas Davis, Daryl Osby, Tom Jordan, Maki Sahara, Megumi Ishimoto, Jackie Steele, Mio Kamitani, and Eric Garcetti.

My greatest professional debts are to the colleagues with whom I worked to create the USGS Multi-Hazards Demonstration Project: Dale Cox, Sue Perry, and Dave Applegate, and to those who have joined me in creating the Dr. Lucy Jones Center for Science and Society to support policy makers in using disaster science: John Bwarie, Kate Long, and Ines Pearce.

My greatest thanks are to my husband, Egill Hauksson, my partner personally and professionally for thirty-seven years, and the foundation of my life.

NOTES

Introduction: Imagine America Without Los Angeles

4 a project we called ShakeOut: Jones et al., *ShakeOut Scenario*.

8 a magnitude 6.2 earthquake occurred in Christchurch: As reported by the New Zealand Parliament, Parliamentary Library Research Paper, Economic Effects of the Canterbury Earthquakes (December 2011), https://www.parliament.nz/en/pb/research-papers/document/00PlibCIP051/economic-effects-of-the-canterbury-earthquakes.

9 In a pilot project in Southern California: Lucy Jones, Richard Bernknopf, Susan Cannon, Dale A. Cox, Len Gaydos, Jon Keeley, Monica Kohler, et al., *Increasing Resiliency to Natural Hazards—A Strategic Plan for the Multi-Hazards Demonstration Project in Southern California*, U.S. Geological Survey Open-file Report 2007-1255, 2007, http://pubs.er.usgs.gov/publication/ofr20071255.

Chapter 1: Brimstone and Fire from out of Heaven

15 The first known product brand: John Day, "Agriculture in the Life of Pompeii," in *Yale Classical Studies*, vol. 3, ed. Austin Harmon (New Haven, CT: Yale University Press, 1932), 167–208.

17 his thirty-seven-volume *Naturalis Historiae*: Pliny the Elder, *Complete Works*, trans. John Bostock (Hastings, East Sussex, UK: Delphi Publishing, Ltd., 2015).

18 "I cannot give you a more exact": Pliny the Younger, *Letters*.

19 Pliny replied that: Pliny the Younger, *Letters*.

21 "Many besought the aid": Pliny the Younger, *Letters*.

22 They are so hot: U.S. Geological Survey, "Pyroclastic Flows Move Fast

and Destroy Everything in Their Path," https://volcanoes.usgs.gov /vhp/pyroclastic_flows.html.

24 St. Augustine of Hippo: Augustine, *Confessions,* trans. H. Chadwick (Oxford: Oxford University Press, 1991).

24 St. Thomas Aquinas: St. Thomas Aquinas, *The Summa Theologica,* trans. Fathers of the English Dominican Province (New York: Benziger Bros., 1947).

Chapter 2: Bury the Dead and Feed the Living

28 "His handsome face": H. Morse Stephens, *The Story of Portugal* (London: T. Fisher Unwin, 1891), 355.

29 "In these affairs": John Smith Athelstane, Conde da Carnota, *The Marquis of Pombal* (London: Longmans, Green, Reader and Dyer, 1871), 28.

32 "[The] table I was writing on": Fordham University, "Modern History Sourcebook: Rev. Charles Davy: The Earthquake at Lisbon, 1755," https://sourcebooks.fordham.edu/mod/1755lisbonquake.asp. From Eva March Tappan, ed., *The World's Story: A History of the World in Story, Song, and Art,* vol. 5, *Italy, France, Spain, and Portugal* (Boston: Houghton Mifflin, 1914), 618–28.

38 "From that day onward": Judith Shklar, *Faces of Injustice* (New Haven, CT: Yale University Press, 1990), 51.

38 "the beginning of a modern distinction": Susan Neiman, *Evil in Modern Thought* (Princeton, NJ: Princeton University Press, 2004), 39.

38 The Lisbon earthquake was not: An analysis of the conflicting opinions is given by Ryan Nichols, "Re-evaluating the Effects of the 1755 Lisbon Earthquake on Eighteenth-Century Minds: How Cognitive Science of Religion Improves Intellectual History with Hypothesis Testing Methods," *Journal of the American Academy of Religion* 82, no. 4 (December 2014): 970–1009.

38 *"What crime, what sin":* Voltaire (François-Marie Arouet), "Poem on the Lisbon Disaster," in *Selected Works of Voltaire,* trans. Joseph McCabe (London: Watts, 1948), https://en.wikisource.org/wiki/Toleration_and_other_essays/Poem_on_the_Lisbon_Disaster.

41 Father Gabriel Malagrida: Kenneth Maxwell, "The Jesuit and the Jew," *ReVista: Harvard Review of Latin America,* "Natural Diasters: Coping with Calamity" (Winter 2007). https://revista.drclas.harvard.edu /book/jesuit-and-jew.

41 "After the earthquake had destroyed": Voltaire (Francoise-Marie Arouet), *Candide* (New York: Boni and Liveright, Inc., 1918), http://www.gutenberg.org/files/19942/19942-h/19942-h.htm.

41 "And what shall we say": John Wesley, *Serious Thoughts Occasioned by the Late Earthquake at Lisbon* (Dublin, 1756).

Chapter 3: The Greatest Catastrophe

46 Irish tradition says that St. Brendan: Katherine Scherman, *Daughter of Fire: A Portrait of Iceland* (Boston: Little, Brown and Co., 1976), 71.

51 "This past week": Jon Steingrimsson, *Fires of the Earth: The Laki Eruption, 1783–1784*, trans. Keneva Kunz (Reykjavík: University of Iceland Press, 1998).

52 "pray to God in correct piety": Alexandra Witze and Jeff Kanipe, *Island on Fire* (New York: Penguin Books, 2014), 87.

52 Even today, Icelandic farmers: Witze and Kanipe, *Island on Fire*, 174.

56 "Such multitudes are": Witze and Kanipe, *Island on Fire*, 120.

Chapter 4: What We Forget

61 When California was lost: Sherburne F. Cook, *The Population of the California Indians, 1769–1970* (Berkeley: University of California Press, 1976).

62 The legislature cut his budget: A Brief History of the California Geological Survey, http://www.conservation.ca.gov/cgs/cgs_history.

64 "Since November 6": William H. Brewer, *Up and Down California in 1860–1864*, ed. Francis Farquhar (New Haven, CT: Yale University Press, 1930), book 3, chapter 1.

65 In Southern California, even less data: W. L. Taylor and R. W. Taylor, *The Great California Flood of 1862* (The Fortnightly Club of Redlands, California, 2007), http://www.redlandsfortnightly.org/papers/Taylor06.htm.

68 both William Brewer: Brewer, *Up and Down California*, book 4, chapter 8.

68 *The New York Times*: "Decrease of Population in California," *New York Times*, October 17, 1863, http://www.nytimes.com/1863/10/17/news/decrease-of-population-in-california.html.

Chapter 5: Finding Faults

77 Japanese mythology holds that Namazu: David Bressan, "Namazu the Earthshaker," *Scientific American,* March 10, 2012, https://blogs.scienti ficamerican.com/history-of-geology/namazu-the-earthshaker/.

78 In the Warring States Period: Joseph Needham, *Science and Civilisation in China,* vol. 2, *History of Scientific Thought* (Cambridge: Cambridge University Press, 1956).

79 described a world: Haiming Wen, *Chinese Philosophy* (Cambridge: Cambridge University Press, 2010), 71.

79 These views were so completely: W. T. De Bary, *Sources of Japanese Tradition,* vol. 1 (New York: Columbia University Press, 2001), 68.

80 "divergence between cosmic moral principles": Gregory Smits, "Shaking Up Japan," in *Journal of Social History* (Summer 2006): 1045–78.

82 Subduction zones such as these: Cliff Frohlich and Laura Reiser Wetzel, "Comparison of Seismic Moment Release Rates Along Different Types of Plate Boundaries," *Geophysics Journal International* 171, no. 2 (2007): 909–20.

89 As his building shuddered: Joshua Hammer, *Yokohama Burning* (New York: Simon and Schuster 2006), 86.

91 These were populist attacks: Smits, "Shaking Up Japan."

93 "the mob lined up children": Sonia Ryang, "The Great Kanto Earthquake and the Massacre of Koreans in 1923: Notes on Japan's Modern National Sovereignty," *Anthropological Quarterly* 76, no. 4 (Autumn 2003): 731–48.

Chapter 6: When the Levee Breaks

97 written by Inca Garcilaso de la Vega: De la Vega, *L'Inca Garcilaso, Historia de la Florida* (Paris: Chez Jean Musier Libraire, 1711), http://international.loc.gov/cgi-bin/query/r?intldl/ascfrbib:@OR (@field(NUMBER+@od2(rbfr+1002))).

100 historian John Barry, in his book: John Barry, *Rising Tide: The Great Mississippi Flood of 1927 and How It Changed America* (New York: Simon and Schuster, 2007), 547.

100 In spite of his own report: A. A. Humphries and Henry L. Abbot, "Report upon the physics and hydraulics of the Mississippi River; upon the protection of the alluvial region against overflow: and upon the deepening of the mouths: based upon surveys and investigations

made under the acts of Congress directing the topographical and hydrographical survey of the delta of the Mississippi River, with such investigations as might lead to determine the most practicable plan for securing it from inundation, and the best mode of deepening the channels at the mouths of the river" (Washington, DC: Government Printing Office, 1867), https://catalog.hathitrust.org/Record/001514788.

101 "The Mississippi River will always": Mark Twain (Samuel Clemens), *Life on the Mississippi* (Boston: James R. Osgood and Co., 1883).

102 because of the depth of the river: J. D. Rodgers, "Development of the New Orleans Flood Protection System Prior to Hurricane Katrina," in *Journal of Geotechnical and Geoenvironmental Engineering* 134, no. 5 (May 2008).

105 "There was needed": U.S. Army Corps of Engineers, *Annual Report of the Chief of Engineers for 1926* (Washington, DC, 1926), 1793.

107 A quasi-governmental commission was created: Kevin Kosar, *Disaster Response and Appointment of a Recovery Czar: The Executive Branch's Response to the Flood of 1927,* CRS Report for Congress, Congressional Research Service, October 25, 2005, https://fas.org/sgp/crs/misc/RL33126.pdf.

108 "probably was good for anybody": George H. Nash, *The Life of Herbert Hoover,* vol. 1, *The Engineer, 1874–1914* (New York: W. W. Norton and Company, 1996), 292–93.

108 By early 1927, most articles: Barry, *Rising Tide,* 270.

108 and those that did commented: See, for example, Alfred Holman, "Coolidge Popular on Pacific Coast," *New York Times,* February 27, 1927.

109 "I can find no warrant": "Veto of the Texas Seed Bill," Daily Articles by the Mises Institute, August 20, 2009, https://mises.org/library/veto-texas-seed-bill.

109 "aid is given freely": Calvin Coolidge, "Speeches as President (1923–1929): Annual Address to the American Red Cross, 1926," archived by the Calvin Coolidge Presidential Foundation, https://coolidgefoundation.org/resources/speeches-as-president-1923-1929-17/.

110 One African American was shot: Winston Harrington, "Use Troops in Flood Area to Imprison Farm Hands," *Chicago Defender,* May 7, 1927.

111 "feel free to make any": Barry, *Rising Tide,* 382.

111 It documented minor misdeeds: American National Red Cross, Colored Advisory Committee, *The Final Report of the Colored Advisory Commission Appointed to Cooperate with the American National Red*

Cross and President's Committee on Relief Work in the Mississippi Valley Flood Disaster of 1927 (American Red Cross, 1929).

113　"We didn't know the Red Cross": "Flood Victim Exposes Acts of Red Cross," *Chicago Defender,* October 15, 1927.

Chapter 7: Celestial Disharmony

119　The first answer came: Peter Molnar and Paul Tapponier, "Cenozoic Tectonics of Asia: Effects of a Continental Collision," *Science* 189, no. 420 (August 8, 1975): 419–26.

120　The result was catastrophic: Jeffrey Wasserstrom, *China in the 21st Century: What Everyone Needs to Know* (New York: Oxford University Press, 2013).

125　That size earthquake: Wang et al., "Predicting the 1975 Haicheng Earthquake."

125　But the peak of such activity: Q. D. Deng, P. Jiang, L. M. Jones, and P. Molnar, "A Preliminary Analysis of Reported Changes in Ground Water and Anomalous Animal Behavior Before the 4 February 1975 Haicheng Earthquake," in *Earthquake Prediction: An International Review,* Maurice Ewing Series, vol. 4, ed. D. W. Simpson and P. G. Richards (Washington, DC: American Geophysical Union, 1981), 543–65.

125　Many opted to self-evacuate: Wang et al., "Predicting the 1975 Haicheng Earthquake," 770.

126　The final count of victims: Wang et al., "Predicting the 1975 Haicheng Earthquake," 779.

128　Already living in bare subsistence: James Palmer, *Heaven Cracks, Earth Shakes: The Tangshan Earthquake and the Death of Mao's China* (New York: Basic Books, 2012).

130　Dong's admonition to the emperor: Tu Wei-Ming, "The Enlightenment Mentality and the Chinese Intellectual Dilemma," in *Perspectives on Modern China: Four Anniversaries,* ed. Kenneth Lieberthal, Joyce Kallgren, Roderick MacFarquhar, and Frederic Wakeman (London and New York: Routledge, 2016).

131　Many books have been written: A good overview is in Palmer, *Heaven Cracks, Earth Shakes.*

132　And Jiang Qing's indictment emphasized: Ross Terrill, *The White-Boned Demon: A Biography of Madame Mao Zedong* (New York: William Morrow and Co., 1984).

Chapter 8: Disasters Without Borders

137 It took fully nine minutes: Z. Duputel, L. Rivera, H. Kanamori, and G. W. Hayes, "Phase Source Inversion for Moderate to Large Earthquakes (1990–2010)," *Geophysical Journal International* 189, no. 2 (2012): 1125–47.

141 Leupeung, a town of ten thousand: James Meek, "From One End to Another, Leupueng Has Vanished as If It Never Existed," *Guardian,* December 31, 2004, https://www.theguardian.com/world/2005/jan/01 /tsunami2004.jamesmeek.

141 The northernmost of the islands: Betwa Sharma, "Remembering the 2004 Tsunami," *Huffington Post India,* December 26, 2014, http://www .huffingtonpost.in/2014/12/26/tsunami_n_6380984.html.

144 The posters, written in English: K. Sieh, "Sumatran Megathrust Earthquakes: From Science to Saving Lives," *Philosophical Transactions of the Royal Society of London* 364 (2006): 1947–63.

Chapter 9: A Study in Failure

152 Still, the year with the most: Hurricane Research Division, National Oceanic and Atmospheric Administration, "Frequently Asked Questions," http://www.aoml.noaa.gov/hrd/tcfaq/E11.html.

153 This further cemented the outcome: David Woolner, "FDR and the New Deal Response to an Environmental Catastrophe," *The Blog of the Roosevelt Institute,* June 3, 2010, http://rooseveltinstitute.org/fdr-and -new-deal-response-environmental-catastrophe/.

155 They called it Hurricane Pam: Madhu Beriwal, "Hurricanes Pam and Katrina: A Lesson in Disaster Planning," *Natural Hazards Observer,* November 2, 2005.

155 New Orleans, however, has the distinction: Robert Giegengack and Kenneth R. Foster, "Physical Constraints on Reconstructing New Orleans," in *Rebuilding Urban Places After Disaster,* ed. E. L. Birch and S. M. Wachter (Philadelphia: University of Pennsylvania Press, 2006), 13–32.

155 The result of this interplay: American Society of Civil Engineers Hurricane Katrina External Review Panel, *The New Orleans Hurricane Protection System: What Went Wrong and Why* (American Society of Civil Engineers, May 1, 2007).

156 Four of these exercise days: Beriwal, "Hurricanes Pam and Katrina: A Lesson in Disaster Planning."

156 Many of the social and engineering: Madhu Beriwal, "Preparing for a Catastrophe: The Hurricane Pam Exercise," statement before the Senate Homeland Security and Governmental Affairs Committee, January 24, 2006, https://www.hsgac.senate.gov/download/012406beriwal.

156 "That 'perfect storm'": "Chertoff: Katrina Scenario Did Not Exist," *CNN*, September 5, 2005, http://www.cnn.com/2005/US/09/03/katrina .chertoff/.

158 The storm surge that hit: R. Knabb, J. Rhome, and D. Brown, *Tropical Cyclone Report: Hurricane Katrina 23–30 August 2005* (Miami: National Hurricane Center, 2006), available at www.nhc.noaa.

158 The total financial losses: Sun Herald Editorial Board, "Mississippi's Invisible Coast," *Sun Herald* (Mississippi), December 14, 2005, http:// www.sunherald.com/news/local/hurricane-katrina/article36463467 .html.

159 In the first few hours: The White House, "The Federal Response to Hurricane Katrina: Lessons Learned," https://georgewbush-whitehouse .archives.gov/reports/katrina-lessons-learned/index.html.

160 They deemed it uninhabitable: U.S. Department of Health and Human Services, "Secretary's Operations Center Flash Report #6," August 30, 2005, quoted in The White House, "The Federal Response to Hurricane Katrina: Lessons Learned," https://georgewbush-whitehouse.archives .gov/reports/katrina-lessons-learned/index.html.

160 "We pee on the floor": Scott Gold, "Trapped in an Arena of Suffering," *Los Angeles Times,* September 1, 2005, http://articles.latimes.com/2005 /sep/01/nation/na-superdome1/.

161 Louisiana reported 1,464 victims: Carl Bialik, "We Still Don't Know How Many People Died Because of Katrina," *FiveThirtyEight*, August 26, 2015, https://fivethirtyeight.com/features/we-still-dont-know -how-many-people-died-because-of-katrina/.

161 The American Red Cross received: "Despite Huge Katrina Relief, Red Cross Criticized," *NBC News*, September 28, 2005, http://www .nbcnews.com/id/9518677/ns/us_news-katrina_the_long_road_back /t/despite-huge-katrina-relief-red-cross-criticized/#.WWLCzdNuIkg.

161 In a bipartisan committee report: Select Bipartisan Committee to Investigate the Preparation for and Response to Hurricane Katrina, *A Failure of Initiative*, 109th Congress, Report 109-377, February 15, 2006, http://www.congress.gov/109/crpt/hrpt377/CRPT-109hrpt377.pdf.

162 Analysis after Katrina: United States Senate, Committee on Home-

land Security and Governmental Affairs, *Hurricane Katrina: A Nation Still Unprepared,* 109th Congress, Session 2, Special Report 109-322, U.S. Government Printing Office, 2006, https://www.hsgac.senate.gov/download/s-rpt-109-322_hurricane-katrina-a-nation-still-unprepared.

162 "a classic example of officials": Russel L. Honoré, *Survival* (New York: Atria Books, 2009), 103.

162 The city didn't know how: Interview with Daryl Osby, Los Angeles County fire chief, May 8, 2017.

163 Governor Blanco did not understand: Spencer Hsu, Joby Warrick, and Rob Stein, "Documents Highlight Bush-Blanco Standoff," *Washington Post,* December 4, 2005, http://www.washingtonpost.com/wp-dyn/content/article/2005/12/04/AR2005120400963.html.

163 After the publicity surrounding: "New Orleans Police Fire 51 for Desertion," *NBC News,* October 31, 2005, http://www.nbcnews.com/id/9855340/ns/us_news-katrina_the_long_road_back/t/new-orleans-police-fire-desertion/#.WTxrBBP1Akh.

163 A Department of Justice investigation: United States Department of Justice Civil Rights Division, *Investigation of the New Orleans Police Department,* March 16, 2011. https://www.justice.gov/sites/default/files/crt/legacy/2011/03/17/nopd_report.pdf.

164 Mayor Nagin left office: Campbell Robertson, "Nagin Guilty of 20 Counts of Bribery and Fraud," *New York Times,* February 13, 2014, https://www.nytimes.com/2014/02/13/us/nagin-corruption-verdict.html.

164 An investigation in 2013: Jeff Zeleny, "$700 million in Katrina Relief Missing," *ABC News,* April 3, 2013, http://abcnews.go.com/Politics/700-million-katrina-relief-funds-missing-report-shows/story?id=18870482.

165 "Looters take advantage": "Looters Take Advantage of New Orleans Mess," *NBC News,* August 30, 2005, http://www.nbcnews.com/id/9131493/ns/us_news-katrina_the_long_road_back/t/looters-take-advantage-new-orleans-mess/.

165 "Relief workers confront": "Relief Workers Confront 'Urban Warfare,'" *CNN,* September 1, 2005, http://www.cnn.com/2005/WEATHER/09/01/katrina.impact/.

165 Lieutenant General Honoré reported: Honoré, *Survival,* 16.

166 An MSNBC report: "New Orleans Police Officers Cleared of Looting,"

NBC News, March 20, 2006, http://www.nbcnews.com/id/11920811/ns /us_news-katrina_the_long_road_back/t/new-orleans-police-officers -cleared-looting/#.WVlTPhP1Akg.

166 "a clearer picture is emerging": Trymaine Lee, "Rumor to Fact in Tales of Post-Katrina Violence," *New York Times,* August 26, 2010, http:// www.nytimes.com/2010/08/27/us/27racial.html.

166 In a chilling echo: John Burnett, "Evacuees Were Turned Away from Gretna, LA," *National Public Radio,* September 20, 2005, http://www .npr.org/templates/story/story.php?storyId=4855611.

166 "Anything coming up this street": Lee, "Rumor to Fact in Tales of Post-Katrina Violence."

166 His trial, postponed many times: John Simerman, "Nine Years Later, Katrina Shooting Case Delayed Indefinitely," *New Orleans Advocate,* August 14, 2014, http://www.theadvocate.com/new_orleans/news /article_736270ed-87ff-58fa-afa4-9b14702854ec.html.

166 On the Danziger Bridge: "Danziger Bridge Officers Sentenced: 7 to 12 Years for Shooters, Cop in Cover-up Gets 3," *Times-Picayune* (New Orleans), April 21, 2016, http://www.nola.com/crime/index .ssf/2016/04/danziger_bridge_officers_sente.html.

Chapter 10: To Court Disaster

170 "I certainly conceive the winds": Pliny the Elder, *Complete Works,* trans. John Bostock (Hastings, East Sussex, UK: Delphi Publishing, Ltd., 2015), chapter 81.

171 "I have found by my inquiries": Pliny the Elder, *Complete Works,* chapter 82.

172 A study done: P. Gasperini, B. Lolli, and G. Vannucci, "Relative Frequencies of Seismic Main Shocks After Strong Shocks in Italy," *Geophysics Journal International* 207 (October 1, 2016): 150–59.

176 The swarm that began: International Commission on Earthquake Forecasting for Civil Protection, "Operational Earthquake Forecasting, State of Knowledge and Guidelines for Utilization," *Annals of Geophysics* 54, no. 4 (2011).

177 He said that the swarm: Richard A. Kerr, "After the Quake, in Search of the Science—or Even a Good Prediction," *Science* 324, no. 5925 (April 17, 2009): 322.

177 Vans mounted with loudspeakers: Thomas Jordan, "Lessons of L'Aquila,

for Operational Earthquake Forecasting," *Seismological Research Letters* 84, no. 1 (2013).

178 "It's better that there are 100": Jordan, "Lessons of L'Aquila," 5.

179 One said he didn't know: Stephen Hall, "Scientists on Trial: At Fault?" *Nature* 477 (September 14, 2011): 264–69.

179 "My father was afraid": Hall, "Scientists on Trial: At Fault?"

180 The government set up: John Hooper, "Pope Visits Italian Village Hit Hardest by Earthquake," *Guardian,* April 28, 2009.

181 "Distracted by Giuliani's predictions": Jordan, "Lessons of L'Aquila."

182 Over the next three years: Edwin Cartlidge, "Italy's Supreme Court Clears L'Aquila Earthquake Scientists for Good," *Science Magazine,* November 20, 2015, http://www.sciencemag.org/news/2015/11/italy -s-supreme-court-clears-l-aquila-earthquake-scientists-good.

Chapter 11: The Island of Ill Fortune

193 Tide gauges in that section: Japanese Meteorological Agency, *Lessons Learned from the Tsunami Disaster Caused by the 2011 Great East Japan Earthquake and Improvements in JMA's Tsunami Warning System,* October 2013, http://www.data.jma.go.jp/svd/eqev/data/en/tsunami /LessonsLearned_Improvements_brochure.pdf.

197 The plant was built: World Nuclear Association, *Fukushima Accident,* updated April 2017, http://www.world-nuclear.org/information -library/safety-and-security/safety-of-plants/fukushima-accident .aspx.

198 Then an explosion: *Scientific American,* "Fukushima Timeline," https:// www.scientificamerican.com/media/multimedia/0312-fukushima -timeline/.

199 The government warned: "Timeline: Japan Power Plant Crisis," BBC, March 13, 2011. http://www.bbc.com/news/science-environment -12722719.

199 After two weeks, even the tap water: *Scientific American,* "Fukushima Timeline."

200 In those first few months: Mizuho Aoki, "Down but Not Out: Japan's Anti-nuclear Movement Fights to Regain Momentum," *Japan Times,* March 11, 2016, http://www.japantimes.co.jp/news/2016/03/11/national /not-japans-anti-nuclear-movement-fights-regain-momentum /#.WVBl5RP1Akg.

Chapter 12: Resilience by Design

210 A dam almost failed: Kenneth Reich, "'71 Valley Quake a Brush with Catastrophe," *Los Angeles Times*, February 4, 1996, http://articles.latimes.com/1996-02-04/news/mn-32287_1_san-fernando-quake.

214 we had created a short movie: "Preparedness Now, the Great California Shakeout," https://www.youtube.com/watch?v=8Z5ckzem7uA.

214 The story we composed: Suzanne Perry, Dale Cox, Lucile Jones, Richard Bernknopf, James Goltz, Kenneth Hudnut, Dennis Mileti, Daniel Ponti, Keith Porter, Michael Reichle, Hope Seligson, Kimberly Shoaf, Jerry Treiman, and Anne Wein, *The ShakeOut Earthquake Scenario—a Story That Southern Californians Are Writing*, U.S. Geological Survey Circular 1324 and California Geological Survey Special Report 207 (2008), http://pubs.usgs.gov/circ/1324/.

216 "If the Big One strikes": Editorial Board, "The Mayor and Preparing for the Big One," *Los Angeles Downtown News*, December 15, 2014, http://www.ladowntownnews.com/opinion/the-mayor-and-preparing-for-the-big-one/article_24cf801a-824a-11e4-a595-1f0a5bc2e992.html.

223 Whereas only 14 percent: *The World Population Prospects, the 2007 Revision*, United Nations Publications, www.un.org/esa/population/publications/wup2007/2007WUP_Highlights_web.pdf.

BIBLIOGRAPHY

Sources used and resources for learning more about natural disasters.

Barry, John. *Rising Tide: The Great Mississippi Flood of 1927 and How It Changed America*. New York: Simon and Schuster, 2007.

Birch, Eugenie, and Susan Wachter. *Rebuilding Urban Places After Disaster*. Philadephia: University of Pennsylvania Press, 2006.

Brewer, William H. *Up and Down California in 1860–1864*. Edited by Francis Farquhar. New Haven, CT: Yale University Press, 1930. Available online at http://www.yosemite.ca.us/library/up_and_down_california/.

Byock, Jesse. *Viking Age Iceland*. London: Penguin Books, 2001.

Carnota, John Smith Athelstane, Conde da. *The Marquis of Pombal*. London: Longmans, Green, Reader and Dyer, 1871.

Honoré, Russel L. *Survival*. New York: Atria Books, 2009.

Hough, Susan. *Earth Shaking Science: What We Know (and Don't Know) About Earthquakes*. Princeton, NJ: Princeton University Press, 2002.

Jones, Lucile M., Richard Bernknopf, Dale Cox, James Goltz, Kenneth Hudnut, Dennis Mileti, Suzanne Perry, et al. *The ShakeOut Scenario*. U.S. Geological Survey Open-File Report 2008-1150 and California Geological Survey Preliminary Report 25, 2008. http//pubs.usgs.gov /of/2008/1150/.

Jordan, Thomas. "Lessons of L'Aquila for Operational Earthquake Forecasting." *Seismological Research Letters* 84, no. 1 (2013): 4–7.

Meyer, Robert, and Howard Kunreuther. *The Ostrich Paradox: Why We Underprepare for Disasters*. Philadelphia: Wharton Digital Press, 2017.

Mileti, Dennis. *Resilience by Design: A Reassessment of Natural Hazards in the United States*. Washington, DC: Joseph Henry Press, 1999.

National Research Council. *Living on an Active Earth.* Washington, DC: The National Academies Press, 2003.

Palmer, James. *Heaven Cracks, Earth Shakes: The Tangshan Earthquake and the Death of Mao's China.* New York: Basic Books, 2012.

Perry, Suzanne, Dale Cox, Lucile Jones, Richard Bernknopf, James Goltz, Kenneth Hudnut, Dennis Mileti, et al. *The ShakeOut Earthquake Scenario: A Story That Southern Californians Are Writing.* U.S. Geological Survey Circular 1324 and California Geological Survey Special Report 207, 2008. http://pubs.usgs.gov/circ/1324/.

Pliny the Elder. *Complete Works.* Translated by John Bostock. Hastings, East Sussex, UK: Delphi Publishing, Ltd., 2015.

Pliny the Younger. "Letter LXV," *The Harvard Classics,* IX, Part 4. Edited by Charles W. Eliot. New York: Bartleby: 1909.

Porter, Keith, Anne Wein, Charles Alpers, Allan Baez, Patrick L. Barnard, James Carter, Alessandra Corsi, et al. *Overview of the ARkStorm Scenario.* U.S. Geological Survey Open-File Report 2010-1312, 2011.

Scherman, Katherine. *Daughter of Fire: A Portrait of Iceland.* Boston: Little, Brown and Co., 1976.

Steingrimsson, Jon. *Fires of the Earth: The Laki Eruption, 1783–1784.* Translated by Keneva Kunz. Reykjavík: University of Iceland Press, 1998.

Wang, Kelin, Qi-Fu Chen, Shihong Sun, and Andong Wang. "Predicting the 1975 Haicheng Earthquake." *Bulletin of the Seismological Society of America* 96, no. 3 (June 2006): 757–95.

Witze, Alexandra, and Jeff Kanipe. *Island on Fire.* New York: Pegasus Books, 2014.

CREDITS

All the maps were created with GMT (P. Wessel, W. H. F. Smith, R. Scharroo, J. F. Luis, and F. Wobbe, "Generic Mapping Tools: Improved Version Released," *EOS Trans. AGU* 94 [2013]: 409–10). Topographic data is from NOAA (https://data.noaa.gov/dataset/5-minute-gridded-global-relief-data-etopo5). The plate boundaries are from Bird (P. Bird, "An Updated Digital Model of Plate Boundaries," *Geochem. Geophys. Geosyst.* 4, no. 3 [2003]: 1027, doi:10.1029/ 2001GC000252).

ABOUT THE AUTHOR

Dr. Lucy Jones was a seismologist for the U.S. Geological Survey for thirty-three years, most recently as science advisor for risk reduction. A research associate at Caltech, she holds a PhD in geophysics from MIT and a BA in Chinese language and literature from Brown University. She lives in Southern California.